SOLID STATE PHYSICS: AN INTRODUCTION

Ognjen Grkinic

TAP

Toronto Academic Press

SOLID STATE PHYSICS: AN INTRODUCTION

Ognjen Grkinic

Toronto Academic Press
4164 Lakeshore Road
Burlington ON L7L 1A4
Canada
www.tap-books.com

Email: orders@arclereducation.com

ISBN: 978-1-77956-332-3

ABOUT THE AUTHOR

Ognjen Grkinić is a physicist with a profound expertise in applied and computer physics, particularly in data science and plasma physics. He is currently pursuing his PhD at the Faculty of Physics, University of Belgrade, where his research focuses on the behavior of ionized gases and plasma within the Attosecond-Plasmas Lab. Ognjen's academic achievements include a Master's degree in applied and computer physics, where he developed advanced software for processing scientific data, earning a near-perfect GPA of 9.80/10.00. In addition to his research, Ognjen has contributed significantly to education, having taught physics and mathematics at both high school and primary school levels. His passion for teaching is matched by his meticulous approach to scientific inquiry, which is evident in the books and papers he has authored and edited in the field of physics and other natural sciences. Beyond academia, Ognjen Grkinić's diverse experiences in research, education, and professional work underscore his commitment to advancing the field of physics and contributing to the broader scientific community.

Table of Contents

List of Figures

List of Tables

PREFACE

Solid state physics has grown significantly in recent decades, becoming one of the largest independent fields of physics. It has simultaneously grown into many new areas, playing an important role in disciplines previously dominated by engineering and chemical sciences. As a result of this rapid development, no single university lecturer can now be expected to have a thorough understanding of all aspects of this vast subject; similarly, it is impossible to imagine a course that could provide students with a comprehensive understanding of the entire discipline and its numerous applications.

Solid-state physics is a fundamental branch of physics that deals with the study of rigid matter, or solids. It encompasses a wide range of phenomena and provides the foundational knowledge required for understanding and advancing numerous technologies. This book, "Solid-state physics: An Introduction," aims to offer a comprehensive overview of the essential concepts, theories, and applications of solid-state physics.

The book is structured into eight chapters, each focusing on a specific aspect of solid-state physics. Chapter 1 provides an overview of the field, including its definition, history, functions, and various types of solids. Chapter 2 discusses the crystal structure and symmetry of solids. It explores periodic arrays of atoms, lattice translation vectors, crystal structures, and the basis for understanding fundamental lattice types in both two and three dimensions.

Chapter 3 is about chemical bonding in solids, discussing various types of bonds including ionic, covalent, van der Waals, metallic, and hydrogen bonds. Chapter 4 focuses on phonons and the thermal properties of solids. It covers topics such as phonon heat capacity, Planck distribution, normal mode enumeration, and the density of states in one and three dimensions. The chapter also discusses the Debye model, anharmonic crystal interactions, thermal expansion, and thermal conductivity, including phonon gas thermal resistivity and Umklapp processes.

Chapter 5 addresses the absorption of light in solids. It explains the absorption coefficient, free carrier absorption in semiconductors and metals, direct and indirect interband transitions, and the conservation of crystal momentum in optical transitions. Chapter 6 deals with the magnetic properties of atoms, focusing on paramagnetism and diamagnetism.

Chapter 7 explores the electronic band structure of solids. It covers general symmetry properties, the nearly free-electron approximation, the tight-binding approximation, and the density of states. Chapter 8 provides an in-depth look at superconductivity. It covers the discovery and occurrence of superconductivity, the thermal and magnetic properties of superconductors, the thermodynamics of the superconducting transition, specific heat, energy gap, and the BCS theory and various applications of superconductivity.

This book has been written specifically for students and scholars to meet their needs in terms of knowledge and to provide them with a broad understanding of business information systems.

—Author

CHAPTER 1

Introduction to Solid-State Physics

LEARNING OBJECTIVES

After studying this chapter, you will be able to:

- Understand the fundamental physical properties of solids and their relation to electronic structure;
- Explore the historical development and functions of solid-state physics;
- Understand the difference between crystalline and amorphous solids and their structural characteristics;
- Identify and differentiate molecular, covalent-network, ionic, and metallic solids based on bonding and properties;
- Know the classification solids into metals, semiconductors, and insulators based on their electrical conductivity and energy band structure;
- Understand the crystal structure and electronic properties of the solids;
- Understand the diverse applications of solid-state physics in technology and innovation.

KEY TERMS FROM THIS CHAPTER

Advanced materials
Magnetic materials
Optoelectronics
Quantum dots
Solid-state physics
Integrated circuits
Nanotechnology
Photovoltaic cells
Semiconductors
Superconductors

UNIT INTRODUCTION

Solid-state physics is a field of study within physics that focuses on investigating the characteristics and behaviors of solid substances. The field of study centers on comprehending the behavior of atoms and electrons within solid materials, with the goal of explaining their physical characteristics, including electrical conductivity, magnetism, and structural integrity. This field involves both theoretical and experimental methodologies to investigate the atomic and molecular configurations in solids and how these configurations influence the characteristics of the material (Hoffmann, 1987).

Key areas of study in Solid-state physics include:

- Crystal Structure: The study of an atom's systematic geometric spatial arrangement within a solid. Crystal structures affect a material's melting point, hardness, and electrical conductivity, among other qualities.
- Electronic Properties: The study of electron behavior in various substances, resulting in the categorization of materials as conductors, semiconductors, or insulators. This includes the examination of band theory and its impact on electron mobility.
- Magnetic Properties: Magnetism in materials is the subject of study that focuses on understanding its genesis and behavior. One must comprehend several forms of magnetic ordering, including antiferromagnetism, ferromagnetism, and paramagnetism.
- Optical Properties: The examination of a material's interactions with light, including light transmission, absorption, and reflection. This area of study investigates phenomena like luminescence and photoconductivity (Castro et al., 2006).
- Thermal Properties: The investigation of heat transfer through solids, encompassing ideas such as thermal conductivity and phonons. Comprehending these characteristics is essential for creating materials with targeted heat control qualities.

The field of solid-state Physics plays a crucial role in advancing cutting-edge technologies and materials, making a significant impact on a wide range of industries such as nanotechnology, electronics, and materials science. It serves as the theoretical basis for the development and enhancement of superconductors, semiconductors, magnetic materials, and other cutting-edge materials utilized in contemporary technology (Elliott & Franz, 2015). This chapter provides an overview of the basic principles that govern the behavior and properties of solid materials. It starts by providing an introduction to solid-state physics, including its historical background and various applications. Furthermore, the categorization of solids is determined by their structural attributes, such as crystalline and amorphous forms, as well as their varied electrical properties. The module investigates the arrangement of crystals, the basic building blocks known as unit cells, and the electrical properties shown by metals, semiconductors, and insulators. The significance of solid-state physics in developing materials research and industrial innovation is highlighted by its applications in numerous industries (Dubovik & Tugushev, 1990).

1.1. OVERVIEW OF SOLID-STATE PHYSICS

The overview of the solid-state physics is explained in the following sections.

1.1.1. Definition of Solid-State Physics

The field of solid-state physics focuses on understanding the fundamental physical properties of solids through the application of basic physical laws. Similar to the study of the dielectric, electrical, elastic, magnetic, and thermal characteristics of solids. Solid-state physics shares more similarities with chemistry than with other branches of physics, as it delves into the shared characteristics of extensive material classes. The study of solid-state physics explores the relationship between physics properties and electronic structure (de PR Moreira & Illas, 2006).

Condensed Matter Physics includes liquids and non-crystalline solids like glass, which we shall not discuss intimately. During the late thirties and forties, modern Solid State Physics came in light. It had its most extensive expansion with the formation of the transistor, integrated circuits, and microelectronics. (Van Genechten et al., 1987).

1.1.2. Background and History

Solid materials consist of closely packed atoms that interact strongly with one other. These interactions give rise to the physical properties of solids, such as their mechanical characteristics (e.g., hardness and elasticity), thermal conductivity, electrical conductivity, magnetic behavior, and optical qualities. Atoms in a substance can be organized in either a regular, geometric structure (crystalline solids, like metals and water ice) or in an irregular manner (amorphous solids, like window glass).

The majority of solid-state physics, as a comprehensive theory, is primarily concerned with crystals. The primary reason for this is that the periodic arrangement of atoms in a crystal, which is its defining characteristic, enables mathematical modeling. Similarly, crystalline materials frequently possess electrical, magnetic, optical, or mechanical characteristics that can be utilized for engineering applications (Martin, 2015).

There are different types of forces that exist between the atoms in a crystal. As an example, consider a crystal of sodium chloride (common salt), where the crystal consists of ionic sodium and chlorine, connected by ionic bonds. In some cases, the atoms share electrons and create covalent bonds. Electrons in metals are shared throughout the entire crystal lattice due to metallic bonding (Ihm, 1988). Eventually, none of these bonding processes occur in the noble gases. The polarization of the electrical charge cloud on each atom produces van der Waals forces, which hold the noble gases together in solid form. The variations in their bonding are what cause the variations in solid kinds.

For many centuries, scientists have been studying the physical characteristics of solids. However, it wasn't until the 1940s that a distinct field called solid-state physics emerged. This field was established with the creation of the Division of Solid-State Physics (DSSP) within the American Physical Society (Keith & Hoch, 1986).

The DSSP primarily served industrial physicists, and solid-state physics became linked with the technical advancements enabled by research on solids. During the early 1960s, the DSSP had become the most extensive division within the American Physical Society (Stern, 1963).

After World War II, significant communities of solid-state physicists also arose in Europe, particularly in England, Germany, and the Soviet Union. Through extensive research and exploration, solid-state physics has emerged as a significant field in both the United States and Europe. It has made significant contributions to the study of semiconductors, superconductivity, nuclear magnetic resonance, and various other interesting phenomena (Mézard & Parisi, 2000).

In the era of the Cold War, the study of solid-state physics extended beyond just solids. This caused certain physicists in the 1970s and 1980s to establish condensed matter physics, a discipline that focused on shared investigative methods for studying solids, liquids, plasmas, and other complicated forms of matter. Nowadays, solid-state physics is widely recognized as a subfield of condensed matter physics, specifically focusing on the characteristics of solids that possess regular crystal lattices.

1.1.3. Functions of Solid-State Physics

Key functions of solids encompass:

- Exploring Crystal Structure: Examining the organization and imperfections in crystalline materials;
- Exploring Electronic Properties: Examining the behavior of electrons in solids, which is essential for understanding electronic devices;
- Exploring Magnetic Properties: Analyzing the magnetic behavior for potential applications in data storage and spintronics;
- Investigating Optical Properties: Gaining insights into the ways in which materials interact with light, with applications in technologies such as lasers and solar cells;
- Investigating Thermal Properties: Analyzing heat conduction and storage in materials to optimize design efficiency;
- Investigating Transport Phenomena: Analyzing charge and heat transport to develop more efficient devices and materials;
- Advancement in Materials Research: Uncovering and creating materials with tailored properties for diverse applications (Zak, 1967).

1.2. TYPES OF SOLIDS

Solids are considered one of the four fundamental states of matter. When atoms come together in a structured, three-dimensional arrangement, the energy of the system decreases, resulting in the formation of a solid from a liquid or gas.

Solids possess distinct qualities that set them apart from liquids and gasses. For example, all solids have the ability to endure forces that are applied either perpendicular or parallel to a surface. The properties of a solid are determined by the features of its constituent atoms, their organization, and the interatomic forces that exist between them (Kotlyar et al., 1998).

The constituent particles of a solid, regardless of whether they are molecular, ionic, covalent, or metallic, undergo significant attractive forces that maintain their positions. Therefore, when considering solids, redirect the attention from the motions of molecules, atoms, or ions, and instead, highlight their immovable positions in space (Thibault et al., 2010).

A solid can arrange its constituent particles in one of two manners: it can adopt a structured and repetitive three-dimensional pattern called a crystal lattice, so forming a crystalline solid, or it can assemble in unpredictable ways, leading to the formation of an amorphous solid.

1.2.1. Crystalline and Amorphous Solids

1.2.1.1. Amorphous

Amorphous compounds possess two unique properties. When an object is cleaved or fractured, it results in uneven and often curved bits. These pieces do not align in a regular manner, causing indistinct patterns in x-rays (Liu et al., 2018).

Glass is often referred to as an opaque, amorphous substance. When the liquid phase of a material is rapidly cooled, it has the potential to solidify in an amorphous state. However, some substances occur naturally in an amorphous state due to either poor fitting of their constituent parts, preventing the formation of a stable crystalline lattice, or the presence of impurities that disrupt the lattice structure (Narten et al., 1976).

1.2.1.2. Crystalline

Crystals possess distinct internal structures, which result in the formation of flat surfaces, commonly referred to as faces. The faces intersect at angles that are characteristic of the material. Every structure produces a distinct pattern when subjected to x-rays, allowing for the identification of the substance. The unique angles show the consistent and repetitive organization of the individual atoms, molecules, or ions in three-dimensional space, regardless of the crystal's dimensions (Vippagunta et al., 2001).

Repulsive interactions, for example, because an ionic crystal to divide along specific planes, resulting in the formation of new surfaces that intersect at the same angles as the original crystal. The orientations at which the surfaces of a covalent solid intersect are not arbitrary. Instead, they are determined by the arrangement of atoms within the crystal (Figure 1.1).

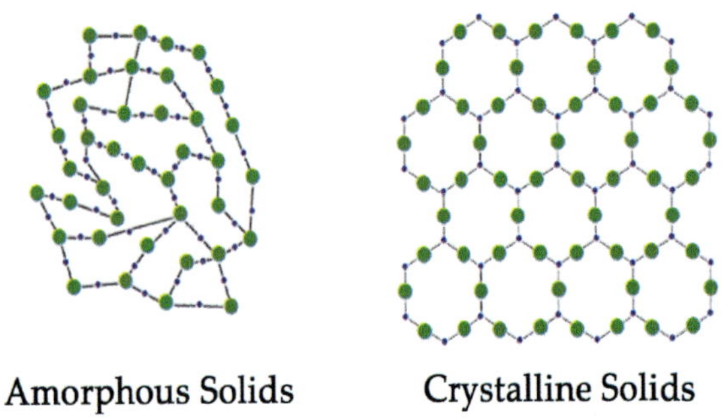

Figure 1.1. Illustration of amorphous and crystalline solids.

Source: geeksforgeek.org, Creative Commons License.

1.2.2. Molecular Solids

Molecular solids are held together by weak intermolecular forces, particularly the Van Der Waals forces, which form fragile crystalline lattices. There is a correlation between the three primary Van Der Waals forces and the three different types of molecular solids. The main concept behind van der Waals forces is the attraction between partial negative and partial positive charges (Seo et al., 2004).

Due to the absence of free electrons, molecular solids composed of neutral molecules lack conductivity. Consequently, electrons exhibit limited conductivity due to their restricted mobility. Because of the weak intermolecular forces, molecular solids exhibit characteristics such as softness, ease of deformation, and low melting points. Furthermore, due to the elongated nature of the molecular bonds, there exists a significant amount of intermolecular space, resulting in solid materials with relatively low density (Mishra et al., 2016).

1.2.2.1. Polar Solids

As the name indicates, the molecules that make up these solids are connected by polarcovalent bonds. The polar nature arises from the variation in electronegativity between the bonded atoms. Therefore, as a result of the variation in electronegativity, atoms acquire partial charges, leading to the formation of dipole-dipole interaction forces that bind the constituent molecules together. As an example, let's consider sulphur dioxide () (Van et al., 1997).

1.2.2.2. Non-Polar Solids

The molecules in these molecular solids are connected or linked together through non-polar bonds. These solids have a soft texture due to the presence of weak Van Der Waals forces that hold their molecules together. The intermolecular bonds are non-polar when they occur between identical atoms or molecules, such as in (a chlorine gas molecule where two chlorine atoms are connected by a single non-polar covalent connection) (Hildebrand,1939).

1.2.2.3. Hydrogen-Bonded Solids

This form of molecular solids exhibits hydrogen bonding, where hydrogen is bound to either fluorine, oxygen, or nitrogen.

Hydrogen fluoride (HF), Water (H_2O), and other similar compounds are examples of molecules that are bonded through hydrogen bonding.

The constituent particles of the substance exhibit polar covalent bonds as a result of the disparity in electronegativity between hydrogen and the other element, which can be either fluorine, oxygen, or nitrogen. Therefore, they possess a rather high level of strength (Somorjai & Hornig, 1962).

1.2.3. Covalent-Network Solids

A solid, such as a crystal or an amorphous solid, is formed when covalent bonds tightly hold its constituent particles together, resulting in a covalent network solid. In a network solid, the atoms form a cohesive and unbroken network. Consequently, the solid can be referred to as a macromolecule since it lacks individual molecules (Yu et al., 2016).

The behavior of covalent network solids is significantly influenced by the covalent bonding present within them.

Some important characteristics to consider are: hardness, elevated melting point, high or low conductivity, and minimal solubility. The hardness of the material is a direct consequence of the strong and resilient covalent bonds it possesses. Due to the inherent difficulty in dissolving network solids, their melting points tend to be quite high. The conductivity of a network solid is influenced by the type of bonding it possesses. Materials that are held together by intermolecular forces demonstrate a higher level of conductivity, in contrast to those that are solely covalently bonded, which exhibit lower conductivity (Chen et al., 2018).

1.2.4. Ionic Solids

Ionic solids are composed of alternating ion lattices. Due to the need for ions to make contact, achieving a high-density packing is not feasible (Dunitz & Orgel, 1960).

The strong electron-ion connection often results in the insulating properties of ionic compounds. These compounds exhibit high density and possess elevated melting temperatures as a result of the robust ionic bonding. Despite their brittleness, the crystal will break if damaged or if it comes into contact with the incorrect ions. Ionic bonding relies on the fundamental principle of lattice energy.

When comparing the electron affinity of a non-metal with the ionization energy of a metal, it becomes evident that the process of generating ions requires the input of energy. When examining the definitions of electron affinities and ionization energies, it becomes apparent that gases are the sole category of substances that fit these criteria. The process of forming ionic solids from the elements is exothermic due to the release of heat during the incorporation of ions into the crystal lattice (Salanne & Madden, 2011).

1.2.5. Metallic Solids

Metallic solids are composed only of metal atoms that are interconnected by metallic bonds. Metallic bonds are formed through the interaction between a lattice of positive ions and delocalized electrons, which provide an intra-molecular force of attraction (Friedel, 1956).

Alkali metals exhibit low melting points, despite the overall tendency of metallic solids to possess high melting temperatures. Each metal possesses a unique melting point when it comes to metallic solids. Metallic substances exhibit exceptional electrical and thermal conductivity due to the easy movement and transmission of their delocalized electrons. Metals are often recognized as materials that can be easily shaped and stretched without breaking, making them ductile and malleable. The shininess of metals can be attributed to the reflection of light by delocalized electrons (Cargill, 1970) (Figure 1.2).

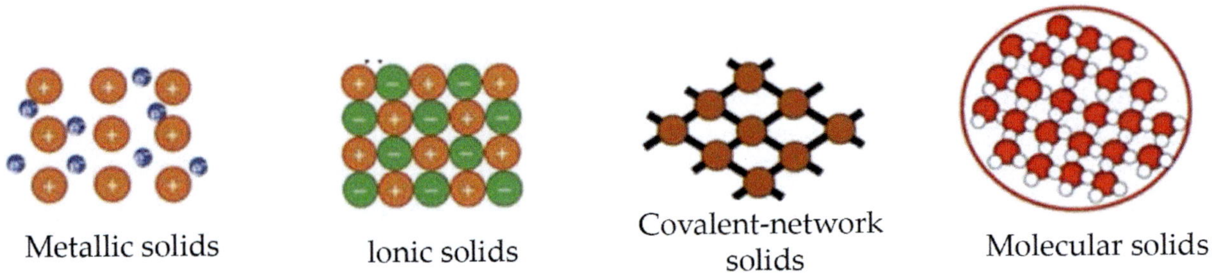

Figure 1.2. Illustration of the different types of solids.

Source: vedantu.com, Creative Commons License.

1.2.6. Metals, Semiconductors, and Insulators

Based on their conductivity or resistivity and energy bands in electronics, solids can be categorized as metals, semiconductors, or insulators (Mott, 1972) (Figure 1.3).

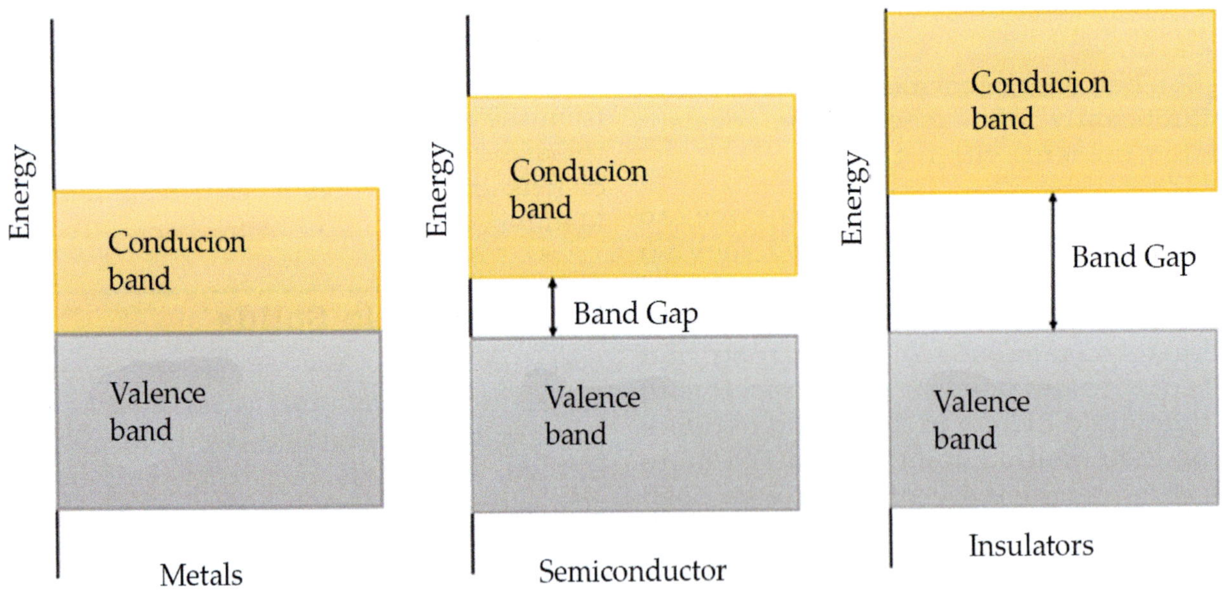

Figure 1.3. Illustration of metals, semiconductor, and insulator.

Source: Harald Ibach, Creative Commons License.

1.2.6.1. Metals

Metals are renowned for their exceptional electrical conductivity, which can be linked to the abundance of unbound electrons. The electrons in this setting are not strongly bound and have the ability to move unrestricted within the metal, which allows for the smooth passage of electric current. In addition to their high thermal conductivity, metals also possess a lustrous appearance and the ability to be easily shaped. The behavior can be explained by the model that describes a "sea of electrons," allowing for the free movement of electrons within a fixed lattice of metal ions (Semrau et al., 2018).

Did you know?

Modern technology was completely transformed by the discovery of semiconductors and their properties. Because of the band gap separating their valence and conduction bands, semiconductors are used in integrated circuits, transistors, and diodes, which are the foundation of contemporary electronics

1.2.6.2. Semiconductors

Electrical conductivity in semiconductors falls between that of conductors (metals) and insulators. These materials possess a distinct characteristic where their conductivity escalates with rising temperature. This phenomenon occurs as an increasing number of electrons acquire sufficient energy to transition from the valence band to the conduction band. Popular semiconductors like silicon and germanium play a vital role in electronic devices by effectively regulating the flow of electrons (Abrams & Wilcoxon, 2005).

1.2.6.3. Insulators

Materials with an extremely high resistance to the flow of electric current are called insulators. In insulators, electrons are firmly bonded to their atoms and must be released with an enormous amount of energy. This significantly slows the flow of electrons across insulators. Examples of materials that are used to cover or enclose conductive materials for safety are rubber, glass, and plastic (Tokura et al., 2019).

1.3. CRYSTAL STRUCTURE AND ELECTRONIC PROPERTIES OF SOLIDS

1.3.1. Crystal Structure and Properties

The crystal structure of materials has a significant impact on a wide range of properties. Various crystallographic techniques, such as X-ray crystallography, electron diffraction, and neutron diffraction, can be employed to analyze this structure.

The dimensions of the individual crystal in a crystalline solid material vary depending on the specific material and the conditions under which it was formed. Many of the crystalline materials commonly encountered in everyday life consist of numerous small crystals, known as polycrystalline structures. Nevertheless, the formation of single macroscopic crystals can occur either in a natural setting (such as gemstones) or in a laboratory environment (Lacorre et al., 1991).

Many of the mechanical and electrical characteristics of actual materials are ultimately determined by the imperfections or abnormalities that authentic crystals display in their flawless structures.

Solid-state physics studies a variety of material characteristics, including heat capacity and electrical conductivity. The Drude model described electrical and thermal conductivity in metals as well as the Hall Effect. It suggested the existence of stationary positive particles and a "gas" of classical, non-interacting electrons. But it greatly exaggerated the electrical heat capacity (Liu et al., 2005).

1.3.2. Crystal Lattices and Unit Cells

Two vectors, a and b, span a two-dimensional lattice. Each point on the lattice can be accessed by a lattice vector that follows the given form:

$$r_n = n_1 a + n_2 b$$

where: and are whole numbers. Various geometries of lattices can be constructed based on the ratio of vector lengths and the angle between them. Figure 1.4 shows the lattice structure that lacks additional symmetry, where and . If every point of this lattice were occupied by an atom, a planar crystal structure would be formed (Rabiei et al., 2020).

Each elementary (or unit) cell with sides a and b would contain one atom, as per the given scenario. These basic units are referred to as primitive. Crystals can sometimes have multiple atoms within their unit cells. In such situations, the lattice points shown in Figure 1.4 represent points within the crystal that share identical surroundings. These points may or may not be located at the center of an atom.

When specific values are assigned to a and b, other planar lattices with higher symmetry can be obtained. When γ equals 90°, the rectangular lattice is obtained as shown in Figure 1.5. Furthermore, when , the lattice transforms into a square configuration. Hexagons can also be arranged in a regular array to fill a two-dimensional plane. The unit cell is determined by the values of a and γ, which are equal to each other and γ is set at 60° (Kirsh & Kupriyanov, 2015).

The unit cell of a hexagonal close packing of spheres is depicted here. When the condition is satisfied, it gives rise to a novel lattice structure. On the other hand, this lattice can be more conveniently described as a "centered" rectangular lattice with unequal sides and a right angle (Figure 1.5). By utilizing an orthogonal coordinate system, one can gain the benefit of a more organized structure. However, it is important to note that the lattice is no longer in its original form (Socolar & Steinhardt, 1986).

Remember

Predicting a solid's characteristics, including its mechanical strength and conductivity, requires an understanding of its crystal structure. The arrangement of atoms in solids is determined by different forms of crystal lattices, which affects the behavior of the materials in a variety of applications, including electronics and building materials.

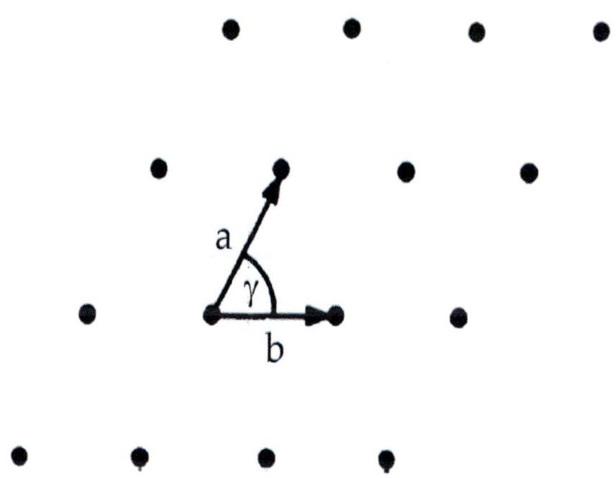

Figure 1.4. Illustration of a plan oblique lattice.

Source: Hans Luth, Creative Commons License.

It becomes evident that the utilization of a centered unit cell is only applicable to a rectangular lattice. When incorporating centering into a square, parallelogram, or hexagonal lattice, it becomes feasible to characterize the resulting lattice using a different set of smaller basis vectors. Put simply, the outcome is that one has achieved an unnecessarily large unit cell (Abarenkov, 2007).

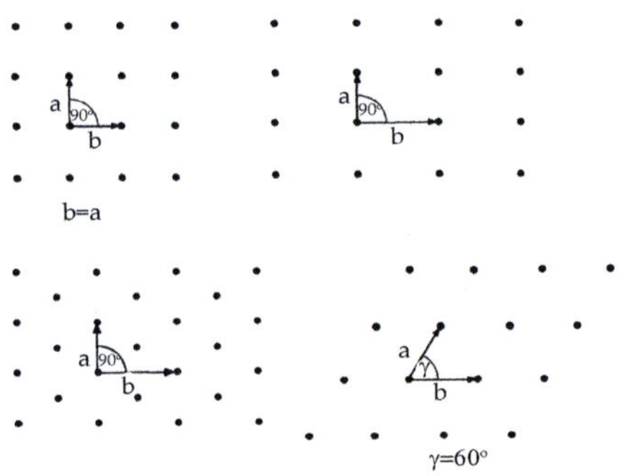

Figure 1.5. Illustration of two-dimensional lattices: square, rectangular, centered rectangular and hexagonal (=60°, a=b).

Source: Huns Luth, Creative Commons License.

The previous discussion has focused on lattices with two dimensions; however, the principles can also be applied to lattices with three dimensions. Instead of the five potential systems of basis vectors in the plane, there are now seven options available (Table 1.1). These relate to the seven unique crystal systems in the field of crystallography. By including the concept of centering, it becomes possible to generate all the potential lattices of three-dimensional space using these fundamental vector systems (Anghel et al., 2002).

In addition to face centering, there is now the option of body centering (Figure 1.6). Just like in two dimensions, it is easy to understand that only specific centerings are practical. Using a tetragonal base-centered lattice would be illogical because it is essentially the same as a primitive tetragonal lattice but with a smaller unit cell (Hayward & Reimers, 1997).

Table 1.1. The Seven Different Basis-Vector Systems or Crystal Systems*

Basis Vectors/ Crystal Axes	Angles	Crystal System
		tetragonal
		monoclinic
		rhombohedral
		orthorhombic
		triclinic
		hexagonal
		cubic

Source: Huns Luth, Creative Commons License.

*Most elements crystallize in a cubic or hexagonal structure. For this reason, and also because of their high symmetry, the cubic and hexagonal coordinate systems are particularly important.

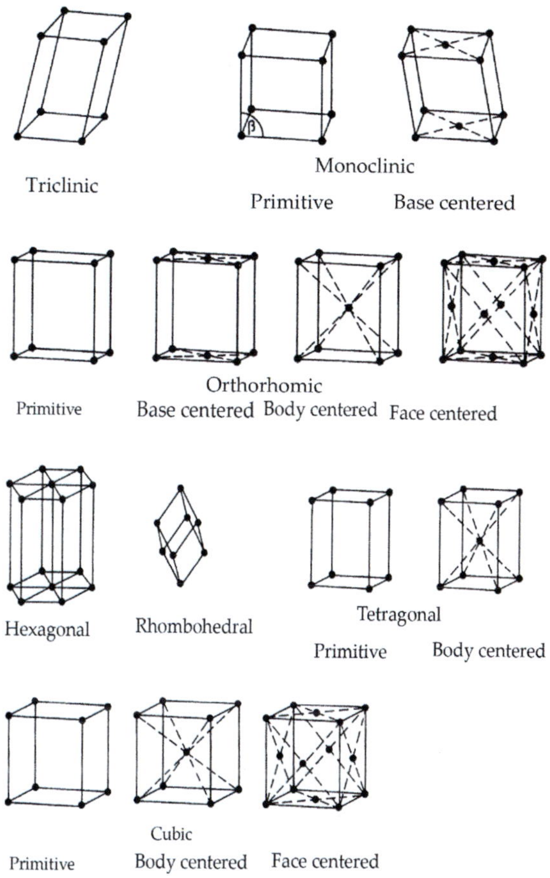

Figure 1.6. Illustration of the 14 three-dimensional Bravais lattices. The hexagonal lattice and the two centered cubic lattices are particularly important in solid-state physics.

Source: Pauling L., Creative Commons License.

1.3.3. Electronic Properties

The study of material attributes like electrical conductivity and heat capacity is known as solid-state physics. Kinetic theory was used by the Drude model, an early framework for electrical conduction, to describe how electrons behaved inside of solids (Ando et al., 1982). The Hall Effect and electrical and thermal conductivity in metals were satisfactorily explained by the Drude model, which postulated the existence of stationary positive ions and an "electron gas" made up of classical, non-interacting electrons. But it significantly overestimated the electrical heat capacity.

Arnold Sommerfeld combined the classical Drude model with quantum mechanics to create the free electron model, also known as the Drude-Sommerfeld model. Within this framework, the electrons are represented as a Fermi gas, a group of particles that follow the principles of quantum mechanics referred to as Fermi-Dirac statistics. While the free electron model made notable progress in predicting the heat capacity of metals, it did not offer a satisfactory explanation for the existence of insulators. The nearly free electron model is a version of the free electron model that mimics the interaction between ions and conduction electrons in a crystalline solid by adding a small periodic perturbation. By presenting the idea of electronic bands, the theory of electronic bands clarifies the presence of conductors, semiconductors, and insulators (Castro et al., 2009). In the nearly free electron model, a periodic potential is taken into account by modifying the Schrödinger equation. Bloch states are the general term for the solutions in this situation. Bloch's theorem can only be used to periodic potentials since atoms in crystals move randomly all the time, disrupting periodicity. As a result, its application is only approximate. Nevertheless, despite this drawback, Bloch's theorem has shown to be incredibly useful. Without it, it would be very difficult to analyze most elements of solid-state physics. Deviations from periodicity are addressed using quantum mechanical perturbation theory (Samarakoon et al., 2011).

1.4. APPLICATIONS OF SOLID-STATE PHYSICS

The field of Solid-State Physics has wide-ranging applications across different industries, driving advancements and fostering innovation. Electronics heavily rely on the creation of semiconductors, transistors, and integrated circuits, which play a crucial role in powering computers, smartphones, and various other electronic devices. Understanding solid-state physics is essential in the advancement of technologies like light-emitting diodes (LEDs) and lasers. These technologies have a wide range of applications, including displays, lighting, and optical communication systems (Andersen et al., 1976).

When it comes to data storage, solid-state physics plays a crucial role in harnessing the power of magnetic materials for hard drives and advancing the technology of magnetic random-access memory (MRAM). This breakthrough leads to faster and more dependable storage options. Renewable energy technologies greatly rely on the principles of solid-state physics to develop high-performance photovoltaic cells for harnessing solar power and thermoelectric materials that efficiently convert heat into electricity (Kerker, 1980).

Medical technology utilizes the principles of solid-state physics in various applications. For example, magnetic resonance imaging (MRI) machines use superconducting magnets, while biosensors are used for health monitoring by detecting biological molecules. Nanotechnology explores the applications of nanomaterials and quantum dots in various fields such as medical imaging, advanced electronics, and solar cells.

Telecommunications heavily relies on the use of optical fibers and solid-state devices to ensure efficient signal transmission and processing, ultimately enhancing global connectivity. Research in solid-state physics is constantly enhancing energy storage technologies, such as advanced lithium-ion , solid-state sodium batteries and super capacitors, resulting in more efficient solutions for storing energy.

Solid-state sensors and actuators, like those found in microelectromechanical systems (MEMS), play a crucial role in a wide range of applications, spanning from automotive to environmental monitoring. Superconductors created through the study of solid-state physics have revolutionized power transmission and enabled the development of groundbreaking technologies such as magnetic levitation, leading to the creation of maglev trains. Research in the field of solid-state physics has greatly contributed to the advancement of display technologies, such as liquid crystal displays (LCDs) and organic light-emitting diodes (OLEDs). These technologies have revolutionized consumer electronics by providing high-resolution screens (Heremans, 1993).

Material science benefits greatly from the field of solid-state physics, as it plays a crucial role in the advancement of high-performance alloys, composites, and ceramics used in various industrial applications. Emerging fields such as quantum computing can greatly benefit from solid-state qubits, offering the potential for much faster processing speeds. Optoelectronics, with its wide range of applications in photo detectors and integrated circuits that combine optical and electronic components, showcases the extensive and influential uses of solid-state physics (Qiu et al., 2020).

ACTIVITY 1.1

Objective: To understand different types of crystal structures found in solids.

Materials Needed:

- Models or digital simulations of different crystal structures (e.g., cubic, tetragonal, hexagonal);
- Paper and pencils for sketching.

Activity Steps:

1. Exploration:
   - Introduce students to models or digital simulations displaying various types of crystal lattices (cubic, tetragonal, hexagonal, etc.).
   - Allow students to examine the models closely, noting the arrangement of atoms or ions within each lattice type.
   - Encourage students to sketch the structures they observe, labeling important components like unit cells and lattice points.
2. Discussion:
   - Gather students for a discussion on their observations.
   - Compare and contrast the different types of crystal lattices in terms of their symmetry, unit cell structure, and potential real-world applications (e.g., metals, semiconductors).
   - Discuss how variations in crystal structure influence material properties such as conductivity, hardness, and optical behavior.

SUMMARY

This chapter offers a comprehensive introduction to the field of solid-state physics, encompassing its historical progression, fundamental principles, and diverse categorizations. The text delves into the evolution of solid-state physics and its importance in comprehending the characteristics of materials.

The chapter provides a comprehensive analysis of solids, classifying them into two main types: crystalline and amorphous. It delves into their atomic structures and thoroughly examines their unique characteristics. Extensive analysis is conducted on crystalline solids, with a specific emphasis on their various lattice structures and the profound impact these structures have on properties such as conductivity and mechanical strength.

In addition, the chapter explores different types of solids based on their molecular bonding, such as molecular solids, covalent-network solids, ionic solids, and metallic solids. Every type is thoroughly examined, taking into account its unique structural arrangements and electronic properties.

The chapter also explores the differences in electronic band structures and conductivity behaviors of metals, semiconductors, and insulators, emphasizing their practical applications in technology and industry.

REVIEW QUESTIONS

1. What are crystalline and amorphous solids? Provide examples and compare their structural characteristics.
2. Explain the bonding types in molecular solids, covalent-network solids, ionic solids, and metallic solids. Give examples of each.
3. Differentiate between metals, semiconductors, and insulators based on their electronic band structures and conductivity properties.?
4. Discuss the significance of crystal structures and unit cells in crystalline solids. Describe common lattice types and their implications for material properties.
5. How does solid-state physics contribute to technological advancements? Provide examples of its applications in electronics and materials science.
6. What role do defects play in the properties of crystalline solids? How do they impact conductivity and mechanical strength?
7. Explain the concept of band gap in semiconductors. How does the size of the band gap affect their conductivity and applications?

MULTIPLE CHOICE QUESTIONS

1. **What type of solids lack long-range order in their atomic arrangement?**
   a. Crystalline solids
   b. Metallic solids
   c. Amorphous solids
   d. Ionic solids
2. **Which type of bonding is characteristic of covalent-network solids?**
   a. Ionic bonds
   b. Metallic bonds
   c. Covalent bonds
   d. Van der Waals forces
3. **Which category of solids is known for having a "sea" of delocalized electrons?**
   a. Molecular solids
   b. Covalent-network solids
   c. Metallic solids
   d. Ionic solids

4 **Semiconductors can be classified as materials that:**
   a. Conduct electricity well
   b. Have a large band gap
   c. Exhibit intermediate conductivity
   d. Lack crystal structure

5. **What is the primary function of solid-state physics?**
   a. Understanding the behavior of gases
   b. Studying the properties of liquids
   c. Analyzing the properties of solids
   d. Exploring the behavior of plasmas

Answers to Multiple Questions

1. (b); 2. (c); 3. (c); 4. (d); 5. (c)

REFERENCES

1. Abarenkov, I. V. (2007). Unit cell for a lattice electrostatic potential. Physical Review B—Condensed Matter and Materials Physics, 76(16), 165127.
2. Abrams, B. L., & Wilcoxon, J. P. (2005). Nanosize semiconductors for photooxidation. Critical reviews in solid state and materials sciences, 30(3), 153–182.

3. Andersen, S. K., Golovchenko, J. A., & Mair, G. (1976). New applications of X-ray standing-wave fields to solid-state physics. Physical Review Letters, 37(17), 1141.
4. Ando, T., Fowler, A. B., & Stern, F. (1982). Electronic properties of two-dimensional systems. Reviews of Modern Physics, 54(2), 437.
5. Anghel, A. T., Day, G. M., & Price, S. L. (2002). A study of the known and hypothetical crystal structures of pyridine: why are there four molecules in the asymmetric unit cell?. CrystEngComm, 4(62), 348–355.
6. Cargill, G. S. (1970). Dense random packing of hard spheres as a structural model for noncrystalline metallic solids. Journal of applied physics, 41(5), 2248–2250.
7. Castro Neto, A. H., Guinea, F., Peres, N. M., Novoselov, K. S., & Geim, A. K. (2009). The electronic properties of graphene. Reviews of modern physics, 81(1), 109–162.
8. Castro, A., Appel, H., Oliveira, M., Rozzi, C. A., Andrade, X., Lorenzen, F., ... & Rubio, A. (2006). basic solid-state physics. phys. stat. sol.(b), 243(11), 2465–2488.
9. Chen, H., Zhang, S., Jiang, W., Zhang, C., Guo, H., Liu, Z., ... & Niu, X. (2018). Prediction of two-dimensional nodal-line semimetals in a carbon nitride covalent network. Journal of Materials Chemistry A, 6(24), 11252–11259.
10. de PR Moreira, I., & Illas, F. (2006). A unified view of the theoretical description of magnetic coupling in molecular chemistry and solid-state physics. Physical Chemistry Chemical Physics, 8(14), 1645–1659.
11. Dubovik, V. M., & Tugushev, V. V. (1990). Toroid moments in electrodynamics and solid-state physics. Physics reports, 187(4), 145–202.
12. Dunitz, J. D., & Orgel, L. E. (1960). Stereochemistry of ionic solids. Advances in Inorganic Chemistry and Radiochemistry, 2, 1–60.
13. Elliott, S. R., & Franz, M. (2015). Colloquium: Majorana fermions in nuclear, particle, and solid-state physics. Reviews of Modern Physics, 87(1), 137–163.
14. Friedel, J. (1956). On some electrical and magnetic properties of metallic solid solutions. Canadian Journal of Physics, 34(12A), 1190–1211.
15. Hayward, J. A., & Reimers, J. R. (1997). Unit cells for the simulation of hexagonal ice. The Journal of chemical physics, 106(4), 1518–1529.
16. Heremans, J. (1993). Solid state magnetic field sensors and applications. Journal of Physics D: Applied Physics, 26(8), 1149.
17. Hildebrand, J. H. (1939). Several Solutions of Non-polar Substances. Journal of Physical Chemistry, 43(1), 109–117.
18. Hoffmann, R. (1987). How chemistry and physics meet in the solid state. Angewandte Chemie International Edition in English, 26(9), 846–878.
19. Ihm, J. (1988). Total energy calculations in solid-state physics. Reports on Progress in Physics, 51(1), 105.
20. Keith, S. T., & Hoch, P. K. (1986). Formation of a research school: Theoretical solid-state physics at Bristol 1930–54. The British Journal for the History of Science, 19(1), 19–44.

21. Kerker, G. P. (1980). Non-singular atomic pseudopotentials for solid state applications. Journal of Physics C: Solid-state physics, 13(9), L189.
22. Kirsh, D., & Kupriyanov, A. (2015). Crystal lattice identification by coordinates of their nodes in three dimensional space. Pattern recognition and image analysis, 25, 456–460.
23. Kotlyar, L. S., Sparks, B. D., Woods, J. R., Raymond, S., Le Page, Y., & Shelfantook, W. (1998). Distribution and types of solids associated with bitumen. Petroleum science and technology, 16(1–2), 1–19.
24. Lacorre, P., Torrance, J. B., Pannetier, J. S. A. I., Nazzal, A. I., Wang, P. W., & Huang, T. C. (1991). Synthesis, crystal structure, and properties of metallic PrNiO3: Comparison with metallic NdNiO3 and semiconducting SmNiO3. Journal of Solid State Chemistry, 91(2), 225–237.
25. Liu, X., Feng, X., Williams, R. O., & Zhang, F. (2018). Characterization of amorphous solid dispersions. Journal of Pharmaceutical Investigation, 48, 19–41.
26. Liu, X., Krott, M., Müller, P., Hu, C., Lueken, H., & Dronskowski, R. (2005). Synthesis, crystal structure, and properties of MnNCN, the first carbodiimide of a magnetic transition metal. Inorganic chemistry, 44(9), 3001–3003.
27. Martin, J. D. (2015). What's in a name change? solid-state physics, condensed matter physics, and materials science. Physics in Perspective, 17, 3–32.
28. Mézard, M., & Parisi, G. (2000). Statistical physics of structural glasses. Journal of Physics: Condensed Matter, 12(29), 6655.
29. Mishra, M. K., Ramamurty, U., & Desiraju, G. R. (2016). Mechanical property design of molecular solids. Current Opinion in Solid State and Materials Science, 20(6), 361–370.
30. Mott, N. F. (1972). The metal-insulator transition in extrinsic semiconductors. Advances in Physics, 21(94), 785–823.
31. Narten, A. H., Venkatesh, C. G., & Rice, S. A. (1976). Diffraction pattern and structure of amorphous solid water at 10 and 77 K. The Journal of Chemical Physics, 64(3), 1106–1121.
32. Qiu, L., Zhu, N., Feng, Y., Michaelides, E. E., ⊠yła, G., Jing, D., ... & Mahian, O. (2020). A review of recent advances in thermophysical properties at the nanoscale: From solid state to colloids. Physics Reports, 843, 1–81.
33. Rabiei, M., Palevicius, A., Dashti, A., Nasiri, S., Monshi, A., Vilkauskas, A., & Janusas, G. (2020). Measurement modulus of elasticity related to the atomic density of planes in unit cell of crystal lattices. Materials, 13(19), 4380.
34. Salanne, M., & Madden, P. A. (2011). Polarization effects in ionic solids and melts. Molecular Physics, 109(19), 2299–2315.
35. Samarakoon, D. K., Chen, Z., Nicolas, C., & Wang, X. Q. (2011). Structural and electronic properties of fluorographene. Small, 7(7), 965–969.
36. Semrau, J. D., DiSpirito, A. A., Gu, W., & Yoon, S. (2018). Metals and methanotrophy. Applied and environmental microbiology, 84(6), e02289–17.

37. Seo, H., Hotta, C., & Fukuyama, H. (2004). Toward systematic understanding of diversity of electronic properties in low-dimensional molecular solids. Chemical reviews, 104(11), 5005–5036.

38. Socolar, J. E., & Steinhardt, P. J. (1986). Quasicrystals. II. Unit-cell configurations. Physical Review B, 34(2), 617.

39. Somorjai, R. L., & Hornig, D. F. (1962). Double-Minimum Potentials in Hydrogen-Bonded Solids. The Journal of Chemical Physics, 36(8), 1980–1987.

40. Stern, F. (1963). Elementary theory of the optical properties of solids. In Solid-state physics (Vol. 15, pp. 299–408). Academic Press.

41. Thibault, J., Alvarez, P. I., Blasco, R., & Vega, R. (2010). Modeling the mean residence time in a rotary dryer for various types of solids. Drying Technology, 28(10), 1136–1141.

42. Tokura, Y., Yasuda, K., & Tsukazaki, A. (2019). Magnetic topological insulators. Nature Reviews Physics, 1(2), 126–143.

43. Van Genechten, K. A., Mortier, W. J., & Geerlings, P. (1987). Intrinsic framework electronegativity: A novel concept in solid state chemistry. The Journal of chemical physics, 86(9), 5063–5071.

44. Van Oss, C. J., Giese, R. F., & Wu, W. (1997). On the predominant electron-donicity of polar solid surfaces. The Journal of Adhesion, 63(1–3), 71–88.

45. Vippagunta, S. R., Brittain, H. G., & Grant, D. J. (2001). Crystalline solids. Advanced drug delivery reviews, 48(1), 3–26.

46. Yu, K., Shi, Q., Li, H., Jabour, J., Yang, H., Dunn, M. L., ... & Qi, H. J. (2016). Interfacial welding of dynamic covalent network polymers. Journal of the Mechanics and Physics of Solids, 94, 1–17.

47. Zak, J. (1967). Finite translations in solid-state physics. Physical Review Letters, 19(24), 1385.

CHAPTER

2

Crystal Structure and Symmetry of Solid

LEARNING OBJECTIVES

After studying this chapter, you will be able to:

- Understand the significance of periodic atomic order in crystalline solids;
- Describe the concept of primitive cells and lattice translation vectors;
- Identify the fundamental types of crystal lattices in two and three dimensions;
- Understand the symmetry operations and constraints defining each lattice type;
- Explain the concept of crystal plane indices and their significance;
- Identify key features and atomic arrangements in sodium chloride and cesium chloride structures;
- Define point symmetry elements: reflection, inversion, and rotation axes;
- Explain how symmetry affects physical properties in crystallography and quantum mechanics.

KEY TERMS FROM THIS CHAPTER

Crystal structure
Index System for crystal planes
Lattice
Primitive cell
Rotation-inversion axes
Crystallographic point groups
Inversion
Point symmetry
Reflection in a plane
Symmetry operations

UNIT INTRODUCTION

When atoms form chemical bonds, they establish specific equilibrium distances based on the principle of minimizing total energy. Thus, a solid with numerous similar atoms has the lowest energy state when each atom is in a homogenous environment. This produces the crystalline state, a three-dimensional periodic pattern. Similarly, this principle applies to solids that consist of multiple elements (Vaida et al., 1988). Here, the periodically repeated units consist of small groups of atoms known as 'building blocks'.

Periodicity in solids leads to certain characteristic features. Periodicity greatly facilitates the theoretical comprehension and formal analysis of solids. While a genuine solid does not exhibit precise three-dimensional periodicity, it is often modeled as having ideal periodicity, and any imperfections are then treated as perturbations (Steed, 2003). Various methods are used to achieve three-dimensional periodic groupings of atoms or "building blocks".

The amorphous state of solids is the opposite of their crystalline state. There is some degree of short-range order present in this situation, but no long-range order. Amorphous solids encompass a variety of materials such as glasses, gels, ceramics, thin-film systems, polymers, and rapidly cooled melts that are formed on a substrate at low temperatures. Research on amorphous materials is one of the busiest fields today.

Despite significant advancements in recent years, there is still much to learn about amorphous materials. One of the reasons is the lack of simplifications that come with periodicity. However, through the analysis of material properties in both crystalline and amorphous states, we have gained valuable insights into the electronic structure and how it influences macroscopic properties. It has been established that short-range order plays a crucial role in determining these characteristics. Therefore, these characteristics exhibit similarities between solids in the amorphous and crystalline state (Cantrell, 1994).

This module delves into the underlying principles that regulate the structure and symmetry of crystalline solids. This chapter explores the concept of unit cells, which are the smallest repeating units that define crystal formations. It highlights the importance of unit cells in predicting material properties. The topics covered in this course include different types of lattices, the use of Miller indices to describe crystallographic planes, and the examination of simple crystal structures such as sodium chloride and diamond (Baur & Tillmanns, 1986). The unit also explores symmetry elements, including rotation axes, inversion centers, and reflection planes, and demonstrates their importance in determining the 32 crystal classes (Nakatsuka et al., 1999).

2.1. PERIODIC ARRAYS OF ATOMS

The discovery of x-ray diffraction in crystals and groundbreaking calculations on crystal characteristics and electron behavior began solid-state physics. What are the advantages of using crystalline solids over non-crystalline solids? Crystals provide the most effective means of expressing the crucial electronic properties of solids (Chakrabarti et al., 2003).

The characteristics of the most essential semiconductors are impacted by the crystalline arrangement of the host material. This is because the short wavelength components of electrons change significantly in response to the material's regular periodic atomic structure. Glasses and other non-crystalline materials are important for optical propagation. This is due to the fact that light waves sense the overall average structure instead of the less uniform local arrangement since they have a longer wavelength than electrons (Gilbert, 1962).

The book begins with crystals. Atoms are added in a steady environment, typically a solution, to create crystals. The initial crystal one may have encountered could have been an authentic quartz crystal that gradually developed over an extended period through the crystallization of a silicate solution in high-temperature water under pressure. The continuous incorporation of identical building components leads to the progressive transformation of the crystal's shape (Barredo et al., 2018).

Here, atoms or collections of atoms serve as the building pieces. Upon its formation, the crystal exhibits a flawless and pure three-dimensional periodic arrangement of identical constituent units, free of any unintended faults or impurities that may have been introduced or integrated into its structure.

Mineralogists discovered that the index numbers, which indicate the orientations of a crystal's faces, are precise integers. This discovery formed the foundation for the initial experimental proof of the periodicity of the crystal's structure. Even further support for the argument was provided by the discovery of x-ray diffraction by crystals in the year 1912. Max von Laue and his colleagues made substantial contributions to the theory of x-ray diffraction by crystals as well as contributions to the experimentation of this phenomenon (Jose et al., 2014).

The importance of x-rays for this task is that they are waves and have a wavelength comparable with the length of a building block of the structure. Such analysis can also be done with neutron diffraction and with electron diffraction, but x-rays are usually the tool of choice.

The diffraction work provided conclusive evidence that crystals consist of a regular arrangement of atoms or groups of atoms. Having a well-defined atomic model of a crystal allowed physicists to explore new horizons, and the emergence of quantum theory played a crucial role in the inception of solid-state physics. Research in this field has expanded to include non-crystalline solids and quantum fluids. Condensed matter physics, a broad and dynamic field, is recognized as one of the largest and most active areas of study in physics (Wang et al., 2022).

2.1.1. Lattice Translation Vectors

The endless repeating of identical atom clusters forms a perfect crystal (Figure 2). A base is an organization made up of people. The term "lattice" refers to the group of mathematical points that are joined to the basis. The three translation vectors, , , and , define the three-dimensional lattice. When observed from the reference point and any other point that is relocated by an integer multiple of the '' vectors, the atomic arrangement in the crystal displays equal symmetry (Barlaud et al., 1994).

$$\mathbf{r}' = \mathbf{r} + u_1\mathbf{a}_1 + u_2\mathbf{a}_2 + u_3\mathbf{a}_3. \quad (2.1)$$

Here, , , can be any integers. The collection of points r' determined by equation (2.1) for all values of , , forms the lattice.

The lattice is considered primitive when the atomic arrangement appears the same for any two points, as long as the integers are chosen appropriately. This statement provides a clear definition of the primitive translation vectors a_i. There is no lower cell volume than . that can be used as a fundamental unit for the crystal structure (Matusov & Ney, 2010). In defining the crystal axes, the primitive translation vectors are frequently used. These vectors constitute three neighboring edges of the primitive parallelepiped. Crystallographers use nonprimitive axes when they match the structure's symmetry (Bilby, 1953) (Figure 2.1).

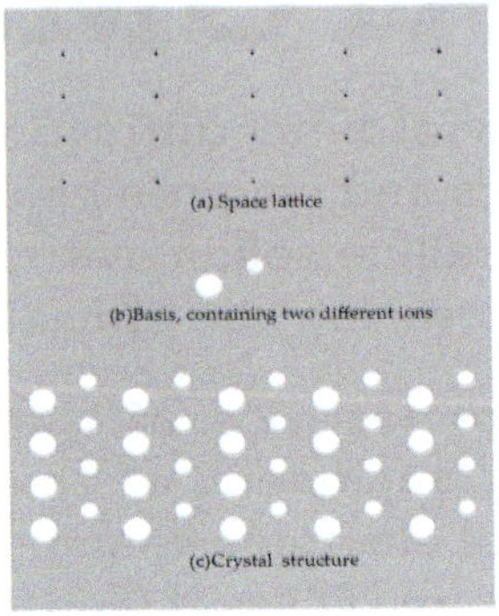

Figure 2.1. Illustration of the crystal structure is created by adding the basis (b) to each lattice point of the space lattice (a). By examining (c), one can identify the fundamental structure and subsequently derive the spatial lattice. The placement of the basis in reference to a lattice point is inconsequential.

Source: Charles Kittel, Creative Commons License.

2.1.2. The Crystal Structure and Basis

Once the crystal axes have been chosen, it is possible to determine the crystal structure. Figure 2.2a shows the procedure of forming a crystal by positioning a basis at each lattice point. It is crucial to understand that lattice points are purely mathematical constructs. All bases within a crystal exhibit similarity in terms of their orientation organization, and composition (Acharya et al., 1994). It is possible for the basis to contain one or more atoms. In relation to the matching lattice point, the position of an atom's center, j, within the basis is:

$$\mathbf{r}_j = x_j\mathbf{a}_1 + y_j\mathbf{a}_2 + z_j\mathbf{a}_3 \ . \quad (2.2)$$

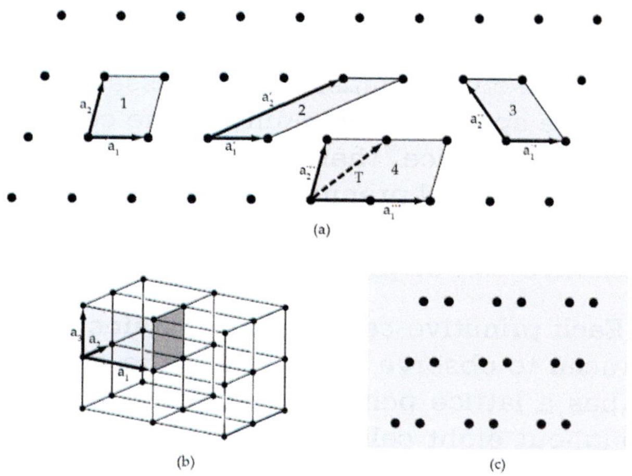

The origin, which we refer to as the associated lattice point, can be arranged so that 0 ≤ , , ≤ 1.

Figure 2.2a. Illustration of (a) the points in two-dimensional spatial lattice where lines met. The lattice translation vectors a_1 and a_2 are both the same. However, since lattice translation T cannot be created by using integral combinations of a1 and a2, a1 and a2 cannot be regarded as primitive translation vectors. The remaining pairs, shown as a1 and a2, can be thought of as the lattice's fundamental translation vectors. Since the areas of the parallelograms 1,2 and 3 are equal, any of them might be considered the original cell. An area of the parallelogram 4 is double that of a primordial cell. (b) Three-dimensional space lattice of a primitive cell (c) Assume that these dots represent the same atoms: Draw a set of lattice points, a primitive cell, a primitive axe selection, and the basis of atoms linked with a lattice point.

Source: K. Urban, Creative Commons License.

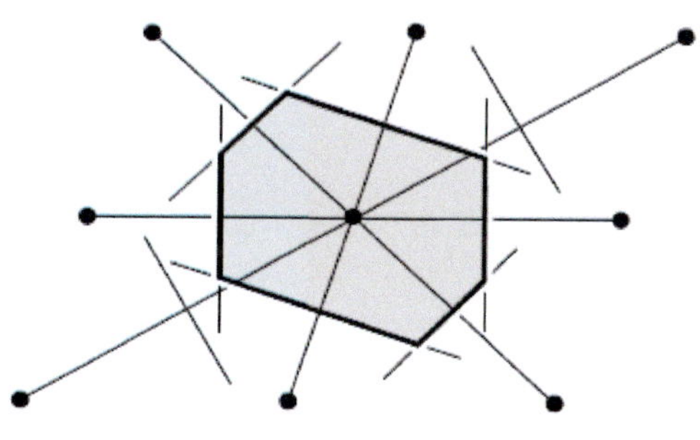

Figure 2.2b. This procedure can help in selecting the primitive cell: (1) Connect a specified lattice point to all adjacent lattice points by drawing lines; (2) Draw new planes or lines at the midpoint and perpendicular to these lines.

Source: T. Nagamiya, Creative Commons License.

2.1.3. Primitive Lattice Cell

A primitive cell is the parallelepiped formed by primitive axes , , (Figure 2.2b). One kind of cell, or unit cell, is a primordial cell. (The adjective unit is unnecessary and unneeded.) By repeating appropriate crystal translation processes, a cell will eventually occupy the entire space. A minimum-volume cell is called a primordial cell. For a given lattice, there are numerous methods for selecting the primitive axis and primitive cell. As long as a given crystal structure is being considered, the number of atoms that are contained within a primitive cell or primitive basis remains constant (Page, 1992).

Remember

The smallest repeating structure that characterizes the overall symmetry and atomic arrangement of a crystal is called its unit cell. Comprehending unit cells facilitates the forecasting and comprehension of crystalline material features.

Each primitive cell always contains exactly one lattice point. It is crucial to observe that each of the eight corners of the primitive cell has a lattice point. These lattice points are distributed evenly throughout eight cells, resulting in each cell containing one lattice point (Lei et al., 2021). This can be calculated by multiplying 8 by 1/8, which equals 1. Simple vector analysis can be used to calculate the volume of a parallelepiped with axes , , . The fundamental structure of a basic cell is known as a primitive basis.

There are always more atoms in a primitive basis than in any other basis. Figure 2.3 shows an alternative method for selecting a primitive cell. Physicists refer to this as a Wigner-Seitz cell.

$$V_c = |\mathbf{a}_1 \cdot \mathbf{a}_2 \times \mathbf{a}_3| \, , \qquad (2.3)$$

2.2. FUNDAMENTAL LATTICES TYPES

Lattice translations T and other symmetry operations can transform or reflect crystal lattices. One common symmetry operation involves rotating around an axis that goes through a point on the lattice. Various lattices exhibit rotation axes that preserve the lattice structure. These axes correspond to rotations by specific angles, such as , , , , and radians, as well as their integral multiples. The symbols 1, 2, 3, 4, and 6 are used to denote the rotation axes (Fisher & Zimmer, 2002).

We have not been able to identify a lattice that exhibits self-similarity under rotations of radians or radians. An individual molecule with a well-designed structure can exhibit any degree of rotational symmetry, while an infinite periodic lattice cannot. It is feasible to generate a crystal by utilizing molecules that individually possess a fivefold rotating axis. However, it should not be anticipated that the crystal lattice itself will also have a fivefold rotation axis. Figure 2.4 demonstrates the outcome of attempting to create a periodic lattice with a pentagonal structure (Reading et al., 2021).

Figure 2.3. Illustration of a period lattice that cannot possess a fivefold axis of symmetry due to the impossibility of completely covering a plane with a connected arrangement of pentagons. It is possible to completely cover a plane with only two different designs of tile or basic polygons.

Source: J.S. Blakemore, Creative Commons License.

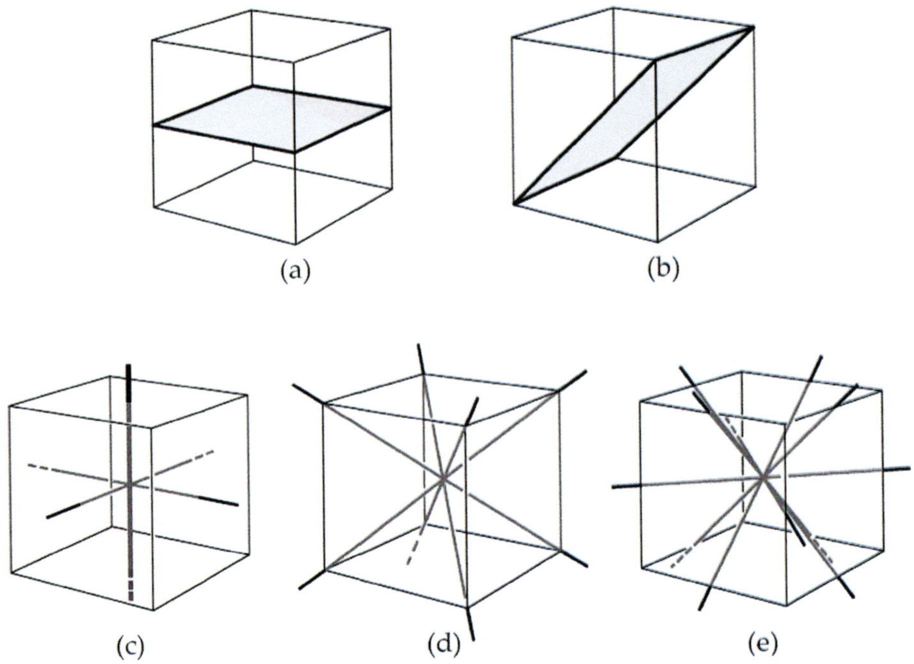

Figure 2.4. Illustration of (a) A symmetry that runs parallel to a cube's faces (b) A cube diagonal plane of symmetry. (c) A cube's three tetrad axes. (d) A cube's four triad axes. (e) A cube's six diad axis.

Source: V.V. Arkadyev, Creative Commons License.

The collection of symmetry operations that, when performed around a lattice point, maintain the lattice's integrity is referred to as the "lattice point group." A long list of possible rotations has been developed. It is possible to have mirror reflections around a plane that is going through a lattice point. The process of inversion involves rotating an object by π radians and then reflecting it in a plane that is perpendicular to the rotational axis (Pan et al., 2018). In the end, r is used in place of -r. The symmetry planes and symmetry axes of a cube are shown in Figure 2.5.

2.2.1. Two-Dimensional Lattice Types

Figure 2.2a shows the lattice drawn for arbitrary values of and . An oblique lattice is a type of lattice that is universally applicable. It remains unchanged when subjected to rotations of π and 2π around any lattice point. However, oblique special lattices can be mirror-reflected or invariant under 2π /3, 2π /4, or 2π /6 rotation (Efremidis et al., 2003).

To build a lattice that is invariant under one or more of these additional operations, we need to impose restrictive requirements on a_1 and a_2. There are four different kinds of restrictions, and each one results in a kind of lattice that we might refer to as special. As a result, in two dimensions, there are five different lattice types: the four special lattices and the oblique lattice depicted in Figure 2.6. The term "Bravais lattice" is commonly used to refer to a particular sort of lattice; in two dimensions, there are five Bravais lattices (Imry & Gunther, 1971).

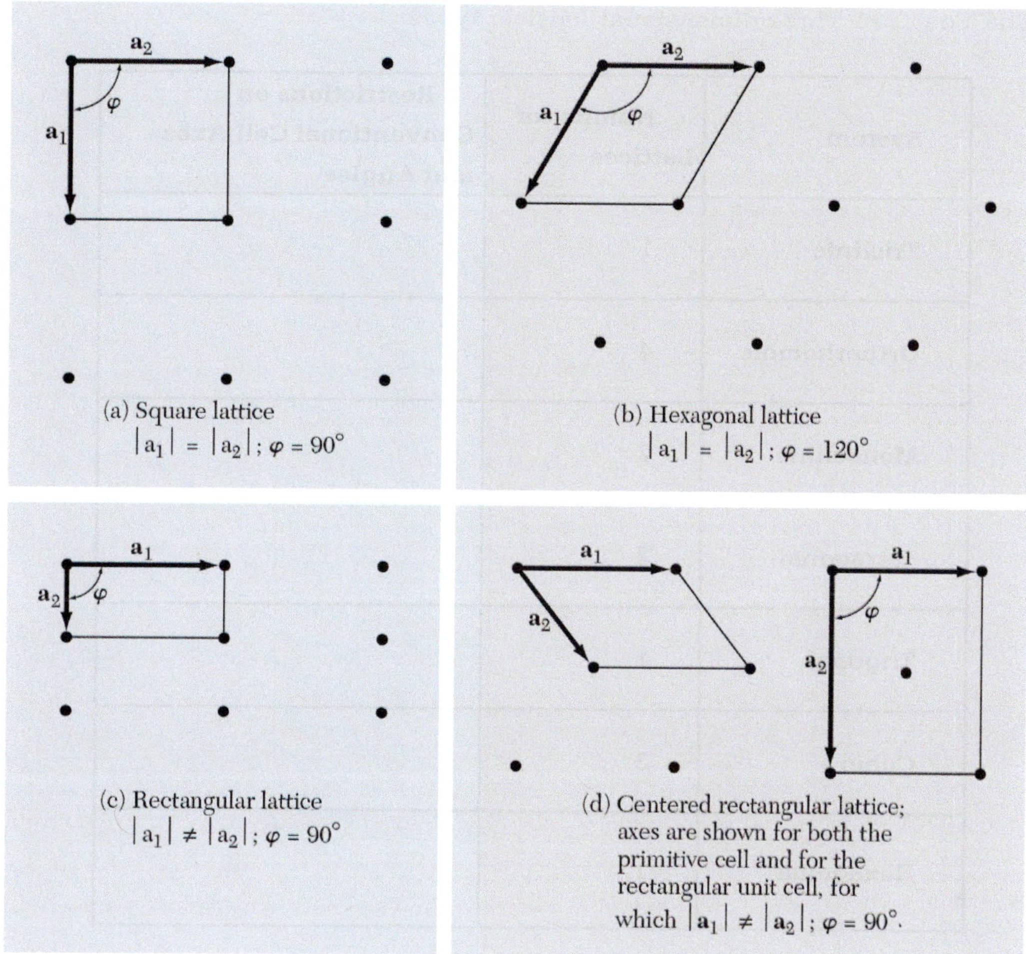

Figure 2.5. Illustration of four special lattices in two dimensions.

Source: S. Foner, Creative Commons License.

2.2.2. Three-Dimensional Lattice Types

Table 2.1 provides the details of the 14 lattice types that are fundamental to three-dimensional point symmetry groups. The overall lattice structure is triclinic, and there are 13 distinct specific lattices.

These systems are conveniently organized into seven types of cells: orthorhombic, triclinic, monoclinic, trigonal, cubic, tetragonal, and hexagonal (Conway & Sloane, 1992). The table expresses the division into systems in terms of the axial relations that characterize the cells. Only the sc is a primitive cell among the typical cells in Figure 2.8 Compared to a primitive cell, a nonprimitive cell frequently exhibits a more direct relationship with the point symmetry operations. The cubic system consists of three lattices: body-centered cubic (bcc), the face-centered cubic (fcc), and simple cubic (sc) lattices (Domb, 1963) (Figure 2.7).

Table 2.1. The Fourteen Three-dimensional Lattice Types

System	Number of Lattices	Restrictions on Conventional Cell Axes and Angles
Triclinic	1	
Orthorhombic	4	
Monoclinic	2	
Tetragonal	2	
Trigonal	1	
Cubic	3	
Hexagonal	1	

Source: Charles Kittel, Creative Commons License.

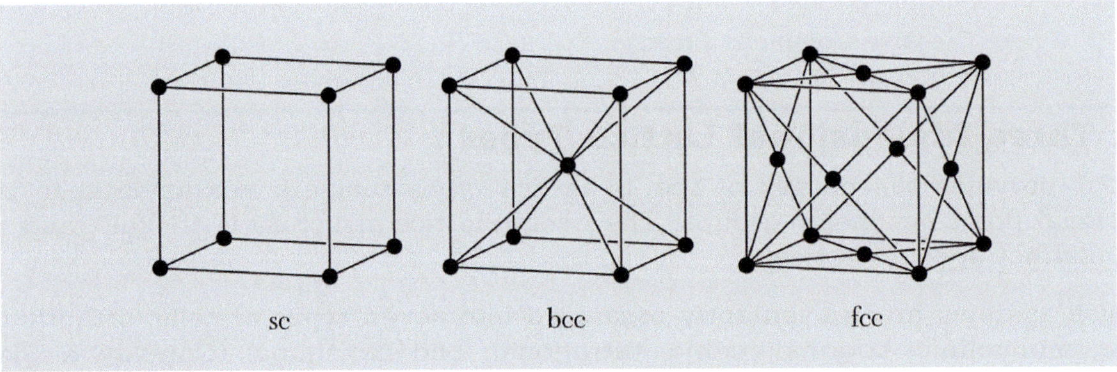

Figure 2.6. Illustration of the lattices of cubic space. The cells that are displayed are the conventional cells.

Source: N. Phillips, Creative Commons License.

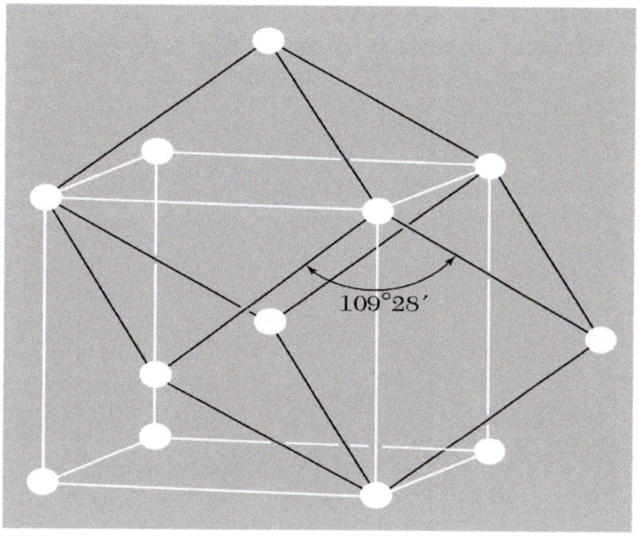

Figure 2.7. Illustration of body-centered cubic lattice that displays a primitive cell.

Source: K.A. Muller, Creative Commons License.

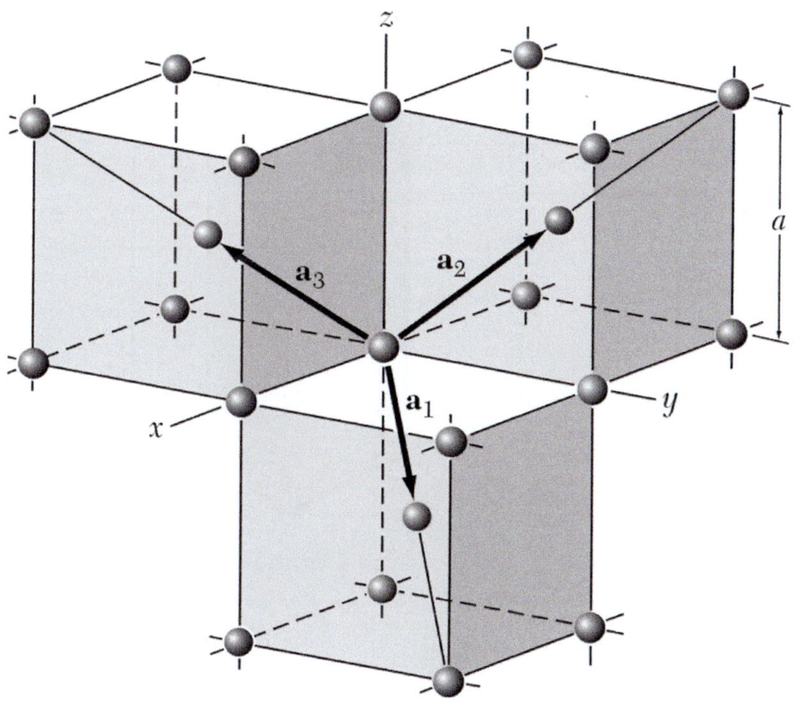

Figure 2.8. Illustration of body-centered lattice translation vectors in primitive form.

Source: R. Ochsenfeld, Creative Commons License.

Figure 2.8 displays a basic cell of the body-centered cubic (bcc) lattice, whereas Figure 2.9 shows the fundamental translation vectors. The fundamental translation vectors of the face-centered cubic (fcc) lattice are shown in Figure 2.10. Primitive cells, by definition, consist of a single lattice point. However, the conventional body-centered cubic (bcc) cell consists of two lattice points, while the face-centered cubic (fcc) cell consists of four lattice points (Feng et al., 2015).

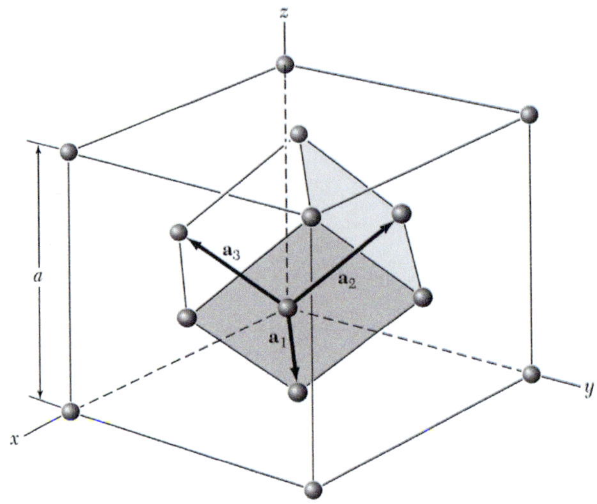

Figure 2.9. Illustration of the face centered cubic crystal's rhombohedral primitive cell.

Source: B.S. Deaver, Creative Commons License.

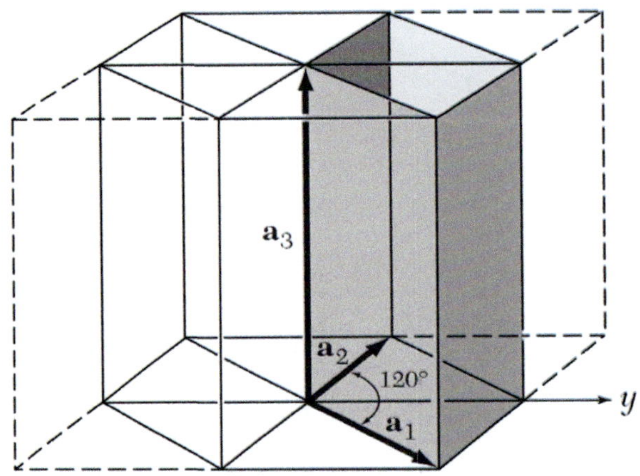

Figure 2.10. Illustration of relationship between the primordial cell and a prism of hexagonal symmetry in the hexagonal system (heavy lines).

Source: D.R. Tilley, Creative Commons License.

The atomic coordinates of x, y, and z determine a point's location within a cell, with (2) defining the position. In this case, the origin is in one corner of the cell, and each coordinate denotes a percentage of the axial lengths , , along the coordinate axis (Kirsh & Kupriyanov, 2015). Therefore, is the coordinate of a cell's body center, and , 011; 101 are its face centers. The primordial cell of the hexagonal system is a right prism with a 120° included angle, based on a rhombus. The link between a hexagonal prism and a rhombic cell is depicted in Figure 2.12 specifies a point's location within a cell in terms of the atomic coordinates x, y, and z. With the origin located at a single cell corner, each coordinate in this case is a fraction of the axial lengths , , in the coordinate axis' direction. Therefore, , is the coordinate of a cell's body center, whereas , 0; . are the coordinates of its face centers. The primordial cell of the hexagonal system is a right prism with a 120° included angle, based on a rhombus (Smardzewski & Wojciechowski, 2019). The link between a hexagonal prism and a rhombic cell is depicted in Figure 2.11.

2.3. INDEX SYSTEM FOR CRYSTAL PLANES

The orientation of a crystal plane is defined by three non-collinear points on the plane. If each point lies on a different crystal axis, the plane can be described by the coordinates of these points in terms of the lattice constants , , . However, it turns out to be more useful for structure analysis to specify the orientation of a plane by the indices determined by the following rules (Lim et al., 2021) (Figure. 2.13).

Determine the axes' intercepts using the lattice constants , , as a guide. The cell's axes could be primitive or nonprimitive;

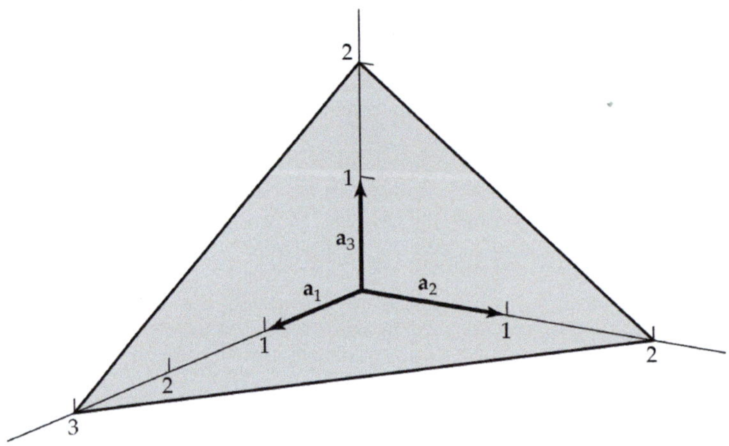

Figure 2.11. The , , axes are intercepted by this plane at , ,2. These numerals have reciprocals of . The indices of the plane are (233) since the smallest three integers with the same ratio are 2,3,3.

Source: L.N. Cooper, Creative Commons License.

For the plane whose intercepts are 4, 1, 2, the reciprocals are , 1, and 1; the smallest three integers having the same ratio are (142). The comparable index for an intercept at infinity is zero. Figure 2.13 shows the indices of a few significant planes in a cubic crystal. A single plane or a group of parallel planes may be indicated by the indices (hkl). A minus sign (hk¯l) is placed above the index to indicate that the associated index is negative if a plane cuts an axis on the negative side of the origin (Liu et al., 2017). A cubic crystal has the following cube faces: (001), (100), (010), (001), (100), (010), and (001).

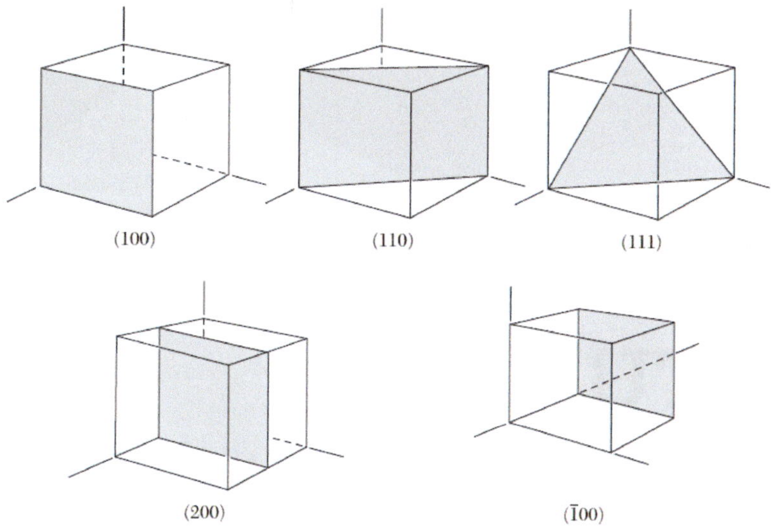

Figure 2.12. Illustration of indicators of a cubic crystal's major planes. The planes (200) and (100) are parallel to one another.

Source: A.B. Pippard, Creative Commons License.

a. Determine reciprocals of numbers, then reduce them to three integers with the same ratio, usually the lowest ratio. It is the result that is represented by the index of the plane, which is symbolized by the symbol (hkl) and is enclosed in parentheses.

The set of cube faces is {100}. Planes equal by symmetry can be indicated by curly brackets (braces) surrounding indices. A plane that cuts the axis at and is parallel to (100) is referred to as the (200) plane.

The indices [] of a direction in a crystal are the set of the smallest integers that have the ratio of the components of a vector in the desired direction, referred to the axes. The axis is the [100] direction; the axis is the [00] direction. In cubic crystals the direction [hkl] is perpendicular to a plane (hkl) having the same indices, but this is not generally true in other crystal systems.

2.4. SIMPLE CRYSTAL STRUCTURES

There is coverage of many common simple crystal shapes, such as those of sodium chloride, diamond, cubic zinc sulfide, cesium chloride, and hexagonal close-packed crystals (Yeh et al., 2004).

2.4.1. Sodium Chloride Structure

The sodium chloride, NaCl, structure is shown in Figs. 2.14 and 2.15. The lattice is face-centered cubic; the basis consists of one Na+ ion and one ion separated by one-half the body diagonal of a unit cube (Klewe & Pedersen, 1974).

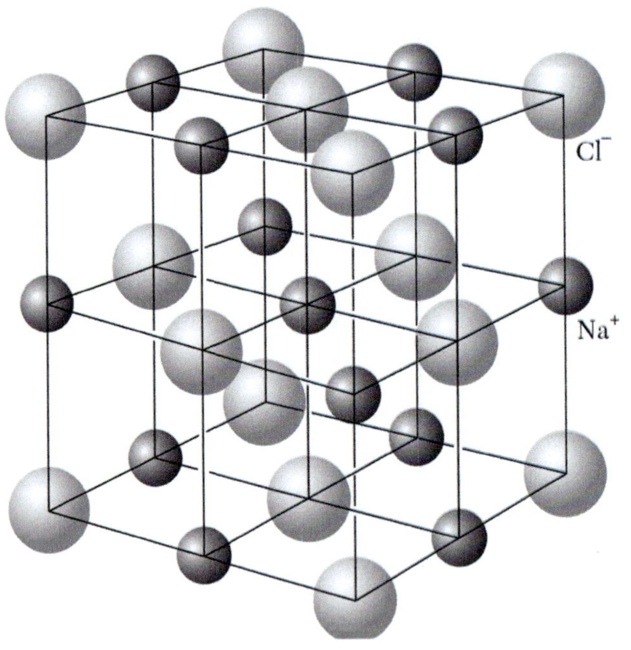

Figure 2.13. Illustration of sodium chloride structure.

Source: M.A. Biondi, Creative Commons License.

Figure 2.14. Illustration of a sodium chloride model; compared to chlorine ions, sodium ions are somewhat smaller.

Source: C.J. Gorter, Creative Commons License.

Figure 2.15. Illustration of lead sulphide (PbS) crystals seen in nature. These crystals have the sodium chloride (NaCl) structure, (photograph by B.Burleson).

Source: F. London, Creative Commons License.

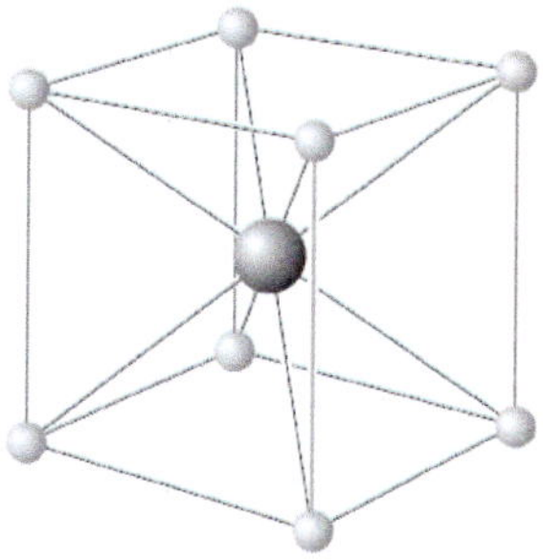

Figure 2.16. Illustration of the cesium chloride crystal structure.

Source: J. Bardeen, Creative Commons License.

Did you know?

The internal symmetry of a crystal allows it to display optical features like birefringence. Spectra of distinct patterns and colors can be observed with polarizing microscopes as a result of birefringence, which happens when light passes through specific crystals and splits into two polarized beams that move at different speeds. Numerous scientific and technological fields rely on this phenomenon, such as materials science, optics, and mineralogy.

Each individual cube has a total of four units of sodium chloride (NaCl), with the atoms organized in specific places (Ismael & Van Fraassen, 2003).

Cl:	000 ;	$\frac{1}{2}\frac{1}{2}0$;	$\frac{1}{2}0\frac{1}{2}$;	$0\frac{1}{2}\frac{1}{2}$.
Na:	$\frac{1}{2}\frac{1}{2}\frac{1}{2}$;	$00\frac{1}{2}$;	$0\frac{1}{2}0$;	$\frac{1}{2}00$.

Each atom is surrounded by six atoms of the opposite type as its closest neighbors. The crystals shown in the table below exemplify the NaCl arrangement. The cube edge, represented as 1 A ⊠, is given in angstroms: 1 . A picture of lead sulfide (PbS) crystals from Joplin, Missouri, is shown in Figure 2.16. The Joplin samples crystallize into exquisite cubes (Belch et al., 1986).

Crystal	*a*	Crystal	*a*
LiH	4.08 Å	AgBr	5.77 Å
MgO	4.20	PbS	5.92
MnO	4.43	KCl	6.29
NaCl	5.63	KBr	6.59

2.4.2. Cesium Chloride Structure

Figure 2.17 depicts the structure of cesium chloride. Each primitive cell consists of a single molecule with atoms located at places 000 and 111 on the simple cubic space lattice, which are the body's centers. It is possible to visualize each atom as the center of a cube containing atoms of the opposite kind, with eight nearest neighbors, or coordination numbers, between them (Tung et al., 2012).

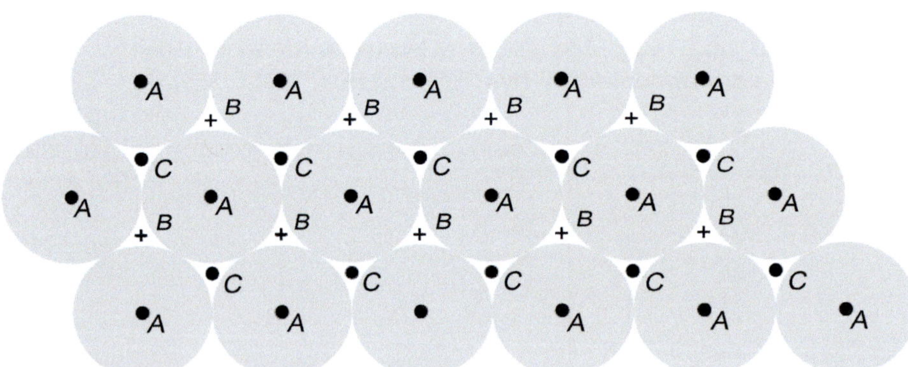

Figure 2.17. Illustration of a spheres dense layer is displayed, with the centers located at position denoted by an A. On top of this, above and parallel to the drawing's plane, is another layer of spheres that is exactly the same, with its center over the spots designated B. A third layer comes in two varieties. It may pass through A or C. The structure is hexagonal close packed and the sequence is ABABAB if it passes over A. The sequence is ABCABCABC and the structure is face centered cubic if the third layer is placed over C.

Source: S. Foner, Creative Commons License.

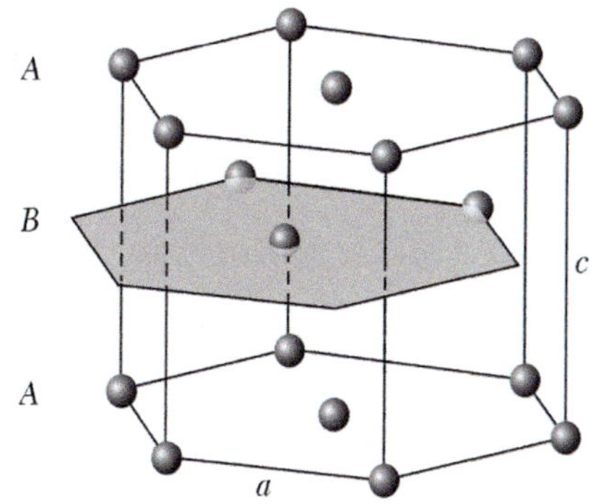

Figure 2.18. Illustration of the hexagonal close-packed structure.

Source: W. Meissner, R., Creative Commons License.

Crystal	a	Crystal	a
BeCu	2.70 Å	LiHg	3.29 Å
AlNi	2.88	NH_4Cl	3.87
CuZn (β-brass)	2.94	TlBr	3.97
CuPd	2.99	CsCl	4.11
AgMg	3.28	TlI	4.20

2.4.3. Hexagonal Close-Packed Structure (HCP)

Identical spheres can be arranged in countless ways to maximize the packing percentage in a regular array (Figure 2.18). There are two types of structures: the hexagonal close-packed structure and the face-centered cubic structure (Figure 2.19). For both constructions, the spheres occupy of the overall volume. Regular or not, no structure has denser packing (Gao et al., 2016).

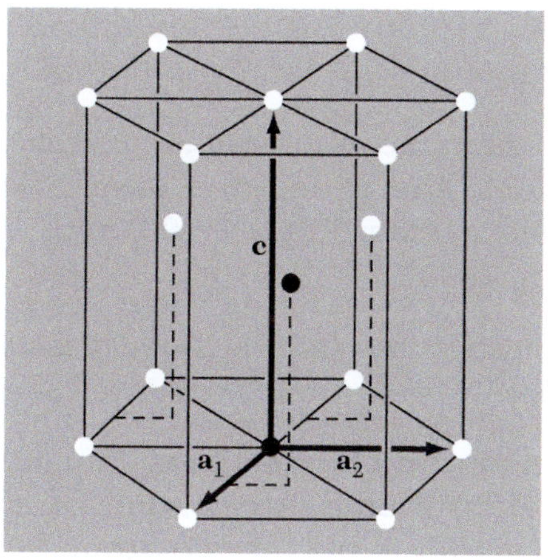

Figure 2.19. Illustration of the primitive cell.

Source: H. Kammerlingh, Creative Commons License.

The closest-packed arrangement of spheres is used to arrange them in a single layer, or layer A, so that every sphere on a plane is in contact with six nearby spheres. This layer can serve as the (111) plane of the face-centered cubic (fcc) structure or the basal plane of a hexagonal close-packed (hcp) structure.

As seen in Figures 2.19 to 2.21, each sphere in layer B is placed in contact with three spheres from the bottom layer to form a second layer, layer B (Zhang et al., 2022).

The third layer, or layer C, can be added in two different ways. The spheres of the third layer are positioned in the first layer's empty regions that B does not fill in the face-centered cubic (fcc) structure.

The spheres in the third layer are precisely positioned so that they are directly above the centers of the spheres in the first layer, forming the hexagonal close-packed (hcp) structure (Partridge, 1967).

Both the hexagonal close-packed (hcp) and face-centered cubic (fcc) configurations consist of twelve atoms in close proximity to one another. There would be no energy difference between the fcc and hcp structures if the binding energy (or free energy) were only based on the number of bonds between nearby atoms (Wei et al., 2019).

Crystal	c/a	Crystal	c/a	Crystal	c/a
He	1.633	Zn	1.861	Zr	1.594
Be	1.581	Cd	1.886	Gd	1.592
Mg	1.623	Co	1.622	Lu	1.586
Ti	1.586	Y	1.570		

2.4.4. Diamond Structure

The diamond structure refers to the distinctive atomic arrangement found in semiconductors like silicon and germanium. This structure is closely linked to the configuration of atoms in other significant binary compounds of semiconductors.

The crystal structure of a diamond is characterized by a face-centered cubic space lattice. According to Figure 2.21, the primordial foundation of the diamond structure consists of two identical atoms at coordinates 000 and connected to every point on the fcc lattice (Bragg & Bragg, 1913).

Considering the face-centered cubic (fcc) lattice, it is obvious that the conventional unit cube contains 4 lattice points.

By extension, since it is made up of two times four atoms, the standard unit cube of the diamond structure has eight atoms. A diamond cannot have its fundamental unit chosen so that the structure is made up of just one atom (Savin et al., 1992) (Figure 2.20).

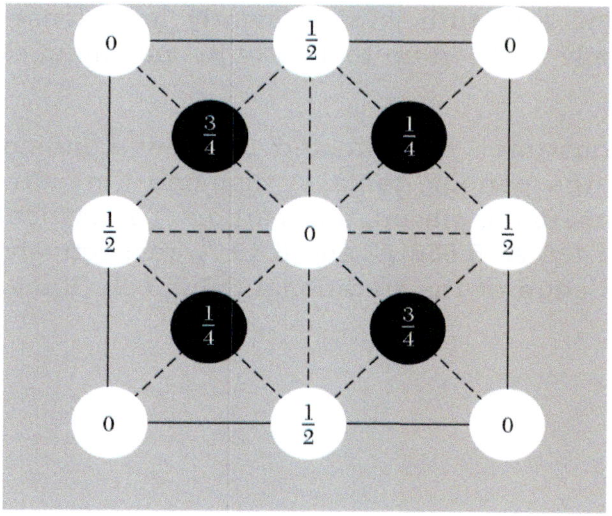

Figure 2.20. Illustration of diamond atomic coordinates as seen through a cube's face; fraction represents the base's height in units of the cube's edge. On the fcc lattice are the points at 0 and 1, while on a similar lattice are the points at 1 and 3.

Source: C. Buzzea, Creative Commons License.

Did you know?

The diamond structure is extremely robust because of its tetrahedral arrangement of carbon atoms. This distinct structure provides a very stiff and strong binding network, making diamond the hardest natural substance known

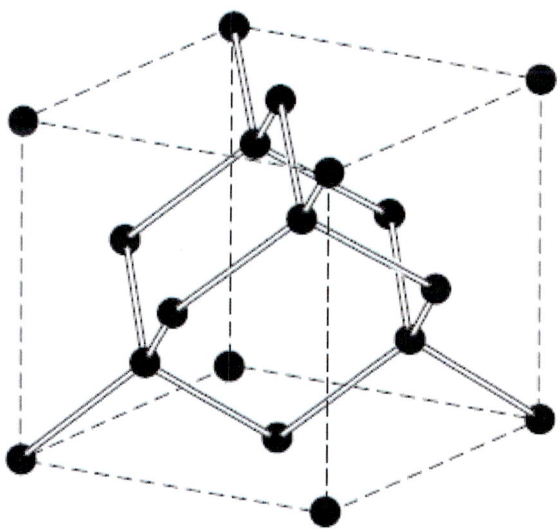

Figure 2.21. Illustration of crystal structure of diamond, showing the tetrahedral bond arrangement.

Source: G. Burns, Creative Commons License.

Figure 2.22 illustrates the diamond structure's tetrahedral bonding feature. Twelve next nearest neighbors and four nearest neighbors exist for each atom. The diamond structure is significantly less dense compared to structures like fcc or hcp. It can only be filled up to 46% of its maximum capacity, which is just 34% (Cowley, 1965).

The diamond structure demonstrates the directed covalent bonding that can be found in elemental column IV of the periodic table. A diamond's crystal structure can be created by combining tin, germanium, silicon, and carbon. The lattice constants for these elements are a = 3.567 Å, 5.430 Å, 5.658 Å, and 6.49 Å, respectively. The variable "a" represents the length of one edge of the traditional cubic cell (Kimball, 1935).

2.5. POINT SYMMETRY

Each point in the lattices we have just described corresponds to an atom or a group of atoms with specific symmetry features. The symmetries and their accompanying notation will be provided in the next section (Bandyopadhyay & Saha, 2007).

2.5.1. Reflection in a Plane

The mathematical expression of this symmetry is achieved through a coordinate transformation. An instance of mirror symmetry with respect to the yz-plane can be described by the transformation where y' is equal to y, z' is equal to z, and x' is equal to the negation of x. In a crystal structure, the presence of a mirror plane is denoted by the sign m. One molecule with two mirror planes that are perpendicular to one another is the water molecule. The locations of the atoms in the molecule define one mirror plane, and the oxygen atom divides the molecule into two equal halves along a second mirror plane that is perpendicular to the first one (Robins, 1991).

2.5.2. Inversion

If $x' = -x$, $y' = -y$, and $z' = -z$ is the coordinate transformation, then inversion symmetry is described by this. As a result, one could say that this qualifies as an act of introspection. is the symbol for inversion symmetry. As seen in Figure 2.23, cyclohexane is an example of a molecule with inversion symmetry. Homo nuclear diatomic molecules have mirror planes and inversion symmetry (Ellis & Oldenburg, 1994).

2.5.3. Rotation Axes

Rotational symmetry is observed when an object remains unchanged after being rotated by a specific angle around a specific axis. One obvious scenario is rotating by 360°, which will always result in the identical structure. The number of intermediate rotations that result in configurations that are indistinguishable is known as the rotation axis order (Hollister et al., 1993).

Therefore, it is possible to have rotation axes of 2-, 3-, 4-, and 6-fold, which correspond to rotations of 180°, 120°, 90°, and 60°, respectively, that result in invariance. Single molecules can also exhibit axes with 5-fold, 7-fold, and other multiples. Clusters composed of small solid aggregates may exhibit 5-fold rotational symmetry. A specific instance is the highly stable icosahedron, which consists of 13 atoms. Icosahedra are created when melts are rapidly cooled (Ter Horst et al., 2010).

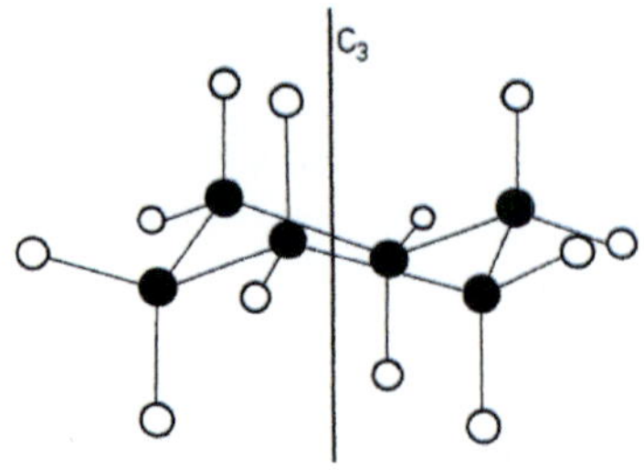

Figure 2.22. Illustration of the molecule of cyclohexane (). The 3-fold axis of rotation is the primary element of symmetry. In addition, the molecule includes three mirror planes, a center of inversion, and three 2-fold rotation axes at 120° perpendicular to the main axis.

Source: M. Tinkham, Creative Commons License.

The resultant solid reflects the localized 5-fold symmetry and has a quasi-crystalline structure that produces unique X-ray diffraction patterns. The only rotation axes that are allowed for strictly periodic crystals are 2-, 3-, 4-, and 6-fold. The necessary translational symmetry is not met by any extra rotations. The numerals 2, 3, 4, and 6 denote the potential axes of rotation (Halvorsen et al., 1999).The cyclohexane molecule seen in Figure 2.23 possesses a 3-fold rotational symmetry axis. Benzene () is a molecule that possesses a 6-fold axis. Its carbon skeleton is composed of a flat and symmetrical hexagon (Hu et al., 2017).

2.5.4. Rotation Inversion Axes

The rotation-inversion axis is a novel symmetry element that can be created by combining rotation with simultaneous inversion. The numbers , , , or stand for it. An axis for threefold rotation and inversion is shown in Figure 2.24. Clearly, a three-fold rotation plus an inversion axis is equal to another three-fold rotation plus an inversion. Combining a mirror plane and a 3-fold rotation axis is an alternative method of expressing the 6-fold rotation-inversion axis (Steinberger & O'Connell, 1997).

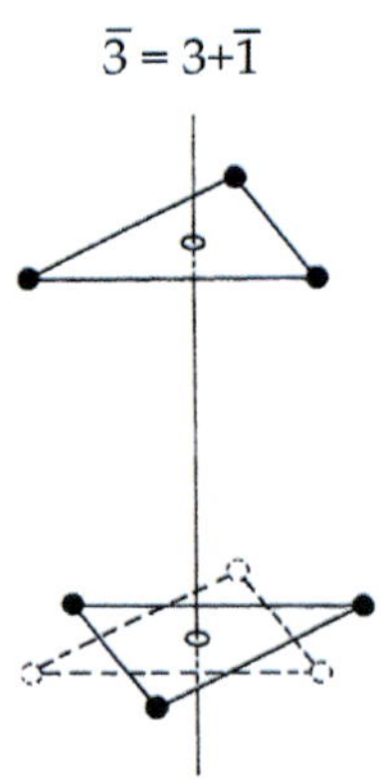

Figure 2.23. Illustration of a threefold axis of rotation and inversion. The combination of other symmetry components can potentially explain the effect.

Source: M. Tinkham, Creative Commons License.

2.6. THE 32 CRYSTAL CLASSES (POINT GROUPS)

There are several ways to combine the symmetry components covered in the previous section. On the other hand, a specific combination of point-symmetry elements can characterize any crystal. The description needs to meet a few requirements in order to be considered complete. For example, if two symmetry operations are conducted consecutively (Kopský, 1979), the outcome needs to be an additional symmetry element: A ⊗ B = C.

According to the associativity rule, the pattern (A ⊗ B) ⊗ C = A ⊗ B ⊗ C must be followed by three or more sequential symmetry operations. The result of the operation A ⊗ E is equal to A since there exists an identity element E, which indicates either no action or a 360° rotation. In addition, each symmetry element A has an inverse , that represents the opposite operation, resulting in A=E (Wilson, 1990).

A group is defined mathematically by these features. 32 different crystallographic point groups are known to exist. The 230 potential space groups are obtained if one further account for translational symmetry. It is important to note that the property of Abelian groups states that for translations A ⊗ B = B ⊗ A, although there may be exceptions to this rule.

The 32 crystallographic point groups are typically depicted using stereographic projections. Crystallographers devised these projections to systematically classify the surfaces of spontaneously grown crystals. Every point where the surface intersects with the sphere is noted and then mapped onto the plane that is perpendicular to the axis of symmetry with the greatest degree of organization (Janssen, 1969).

Points where the plane intersects above are represented by solid circles, while those on the lower half sphere are indicated by empty circles or crosses. When point groups are represented in a systematic manner, the axis that has the highest order is situated in the middle of the depiction. Figure 2.25 illustrates stereographic projections of groupings that consist of two points. There are three alternative ways to represent a specific point group:

- By defining a set of rules for creating symmetry operations;
- By indicating the symbol of international point group;
- By the Schonflies symbol (Mozrzymas, 1974).

When it comes to notation, generating symmetry operators is widely used in crystallography, whereas group theory and spectroscopy have primarily used Schoenflies symbols. The Schoenflies notation comprises a primary symbol representing the axis of rotation, if present, and a secondary symbol indicating the positions of the symmetry planes.

For example, let's examine the water molecule, which has a 2-fold axis as its highest order rotation axis. Vertical orientation of the symmetry planes indicates that they run parallel to the main axis of rotation. The symbol v corresponds to the given Schoenflies symbol (De Wolff, 1974).

There are three fourfold axes of rotation, four threefold axes of rotation, and symmetry planes perpendicular to the fourfold axes It is represented by the symbol in Schoenflies (Harker, 1976).

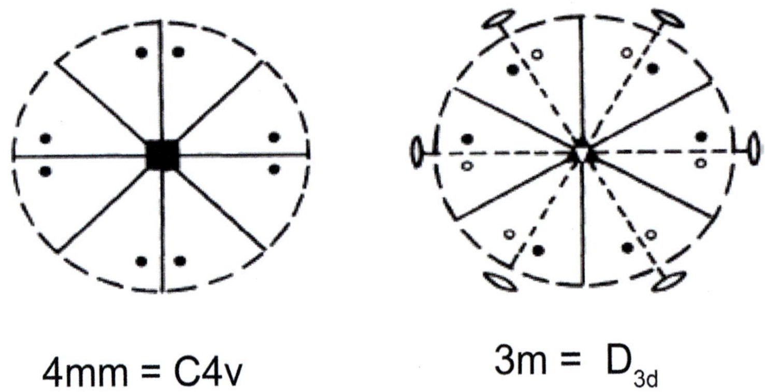

Figure 2.24. Stereographic projection representation of two point's group's symmetry elements.

Source: D. Schoenberg, Creative Commons License.

2.7. THE SIGNIFICANCE OF SYMMETRY

For those who are not familiar with it, the accurate allocation and representation of symmetry can frequently appear complex and confusing. Therefore, it would be beneficial to provide a concise explanation of the significance of symmetry in characterizing a solid. To accomplish this objective, we rely on the principles and concepts of quantum physics as the foundation of our discussion. There are two mirror planes in a water molecule. All of the molecule's physical properties must demonstrate the presence of these two mirror planes (Brading & Brown, 2004).

If a molecule's electrical or vibrational properties can be characterized by a Hamiltonian, it exhibits 2-fold mirror symmetry, meaning it remains unchanged when subjected to the associated coordinate transformation (Forrest, 2015). This invariance can also be articulated using alternative methods. A reflection is associated with an operator σ. When the operator σ applies to the Hamiltonian H, an eigenstate ψ, or R, the result should show H, ψ, or R in the transformed (reflected) coordinates. Matrices are used to represent operators. The matrix operation corresponds to the mirroring of coordinates across the yz-plane (Gabrielse, 1980).

$$\begin{pmatrix} -1 & 0 & 0 \\ 0 & 1 & 0 \\ 0 & 0 & 1 \end{pmatrix} \begin{pmatrix} x \\ y \\ z \end{pmatrix} = \begin{pmatrix} -x \\ y \\ z \end{pmatrix} \tag{2.4}$$

This is a three-dimensional depiction. Additionally, this operation can be represented using three separate one-dimensional matrices, with each matrix specifically affecting one coordinate component. The three-dimensional representation in this case is called "reducible", but the analogous one-dimensional representation is called "irreducible" as it cannot be made any simpler (Caulton, 2015).

It is evident and clear that the irreducible representation of a rotation through 180° (a two-fold rotation axis) has one dimension. Reversing the sign allows for a clear description to be given by selecting appropriate coordinates. But for the 3-, 4-, and ⊠-fold rotation axes, the processes always require two coordinate changes in addition to a full 360° rotation. The irreducible representation can be inferred to be two-dimensional (Mondal et al., 2017).

$$[(-1)x;(1)y;(1)z] = (-x;y;z),$$

When the Hamiltonian operator exhibits a specific symmetry, such as mirror symmetry, the order of the reflection operation and the Hamiltonian operator does not matter. In other words, these two operators commute (Kovesi, 1997).

It is widely recognized in the field of quantum mechanics that these operators possess a shared set of Eigen states. Therefore, the potential Eigen states of H can be categorized based on their eigenvalues in relation to the symmetry operators (Tari, 1999). When considering mirror symmetry and a 2-fold rotation axis , it is important to note that = 1 and = 1. As a result, the eigenvalues are limited to ±1.

$$\sigma\Psi_{+} = +\Psi_{+}, \qquad C_2\Psi_{+} = +\Psi_{+},$$
$$\sigma\Psi_{-} = -\Psi_{-}, \qquad C_2\Psi_{-} = -\Psi_{-}. \qquad (2.5)$$

Accordingly, the Eigen states of H with regard to these operators might be symmetric or anti-symmetric (Preskill et al., 1991). The Eigen states' even or odd "parity" is a common way to describe this. We have already seen an example of both even and odd parity in our investigation of the chemical bonding between hydrogen atoms. As a symmetric collection of atomic wave functions, the bonding state was an even parity state (White & Foulds, 2018).

The Eigen states and correspond to different eigenvalues of H, as seen in this example. This means that the matching energy levels are non-degenerate. For example, this implies that the water molecule is restricted to non-degenerate energy states (we will refrain from discussing normal mode vibrations or incidental degeneracy of energy levels in this context) (Murgueitio, 2022).

Next, we'll apply these concepts to the research of water molecules' normal mode vibrations to support the arguments presented before. As a result, the atoms have the ability to move in a symmetrical or anti-symmetrical manner in relation to the molecule's two mirror planes. When atoms are positioned in a mirror plane, their motion must be anti-symmetric in order for reflection to reverse the motion. This means that the motion will be perpendicular to the plane (Gardner & Davies, 2006).

The atoms must undergo symmetric vibrations that occur within the mirror plane. One of the O molecule's mirror planes, which has the three atoms, is shown in Figure 2.25. The molecule demonstrates anti-symmetry in relation to this plane through two distinct movements: rotations and translations that are perpendicular to the plane (Davi et al., 1992).

There are six symmetric modes that involve motion within the molecule's plane. These modes consist of three vibrations in the normal mode, two translations, and a rotation around a perpendicular axis to the plane. Among these vibrations, one exhibits anti-symmetry in relation to the mirror plane that is perpendicular to the molecular plane, while the other two display symmetry (Tornari, 2022).

The classification of vibrational modes and electronic states can be applied to more complex compounds as well. On the other hand, for operators with two-dimensional irreducible representations, like , the procedure becomes somewhat more complicated. In addition to the state ψ, the state ψ is also an Eigen state of if commutes with H. There are now two options.

- Except for a numerical factor, which may be set to 1 through appropriate normalization, the state T is exactly the same as T. T thus exhibits symmetry with regard to , and the operation ,is represented in one dimension (numerically) in this case. With respect to the operation, the state ψ is not degenerate (Kojić & Štefanić, 2010).
- The application of the operator on the state ψ results in a new state ψ', which is linearly independent. Because and H commute, ψ' also represent an eigenstate of, having the same eigenvalue E (Eady et al., 1995). States ψ and, ψ' are hence degenerate. A two-dimensional matrix can express the irreducible representation of the rotation as it always affects two coordinates. Each Eigen state of can be formed by combining two functions that are chosen to be orthonormal. The energy levels exhibit a 2-fold degeneracy. These degenerate levels are present in all point groups that have a rotation axis with a magnitude greater than 2 (Bates et al., 1988).

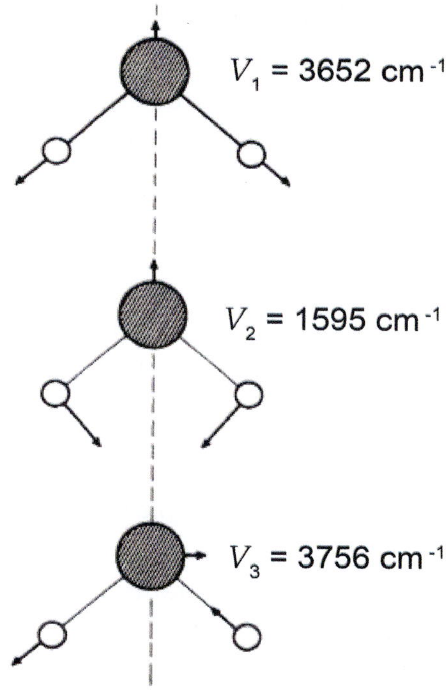

Figure 2.25. Illustration of the water molecule that exhibits two types of vibrations: symmetric and anti-symmetric. Together these three translations and three rotations represents nine normal modes that corresponds to the nine degrees of freedom.

Source: C. Kittel, Creative Commons License.

ACTIVITY 2.1

Objective: To understand and demonstrate crystal symmetry elements.

Materials Needed:

- Pen and paper;
- Access to a crystal structure diagram (from textbook or online).

Steps:

1. Identifying Symmetry Elements:
   - Distribute a crystal structure diagram to each participant or group.
   - Ask participants to carefully study the diagram and identify different symmetry elements such as:
     a. Mirror planes;
     b. Rotation axes;
     c. Inversion centers;
     d. Glide planes (if applicable).
   - Encourage participants to annotate the diagram to mark each identified symmetry element.
2. Discussion and Presentation:
   - Gather participants together for a discussion;
   - Ask each participant or group to present their findings;
   - Describe the symmetry elements they identified;
   - Explain how each symmetry element affects the crystal's structure and properties;
   - Facilitate a discussion on the importance of symmetry in crystallography and its implications for material science and technology.

SUMMARY

- The chapter explore into the fundamental principles governing the arrangement of atoms in crystalline materials. It begins by discussing periodic arrays of atoms, defined by lattice translation vectors that establish the repetitive structure. The concept of a basis is introduced, describing how atoms or groups of atoms are positioned within the unit cell, which is the smallest repeating unit defining the crystal structure.
- It also categorizes crystal lattices into fundamental types based on symmetry and unit cell shape, covering both two-dimensional and three-dimensional arrangements. It explores the Miller indices system for crystallographic planes and examines various simple crystal structures such as sodium chloride and diamond structures, highlighting their atomic arrangements and symmetrical properties.
- The discussion then moves to symmetry operations, including point symmetry elements such as reflection, inversion, rotation axes, and rotation inversion axes. These elements are crucial in defining the 32 crystal classes, or point groups, which classify crystals based on their symmetry properties. Finally, the chapter underscores the significance of symmetry in crystals.

REVIEW QUESTIONS

1. Describe the concept of a unit cell in crystallography. Why is it important in understanding crystal structures?
2. Compare and contrast primitive lattice cells with non-primitive (conventional) lattice cells. Provide examples of each.
3. Explain the significance of Miller indices in crystallography. How are Miller indices determined for a crystal plane?
4. Discuss the differences between the sodium chloride (NaCl) structure and the diamond structure in terms of their atomic arrangements and symmetry.
5. What are point symmetry operations in crystals? Provide examples of such operations and their effects on crystal symmetry.
6. Define rotation axes and rotation-inversion axes in crystallography. Give examples of crystals that exhibit these symmetries.
7. Briefly explain how crystallographic symmetry influences the physical properties of solids such as electrical conductivity and optical behavior.
8. What are the 32 crystal classes (point groups)? How are these classes determined, and what role do they play in crystallography?
9. Describe the role of inversion symmetry in crystal structures. How does inversion symmetry affect the arrangement of atoms within a crystal?
10. Discuss the classification of crystal structures based on lattice types. Provide examples of different lattice types and their characteristic features.

MULTIPLE CHOICE QUESTIONS

1. **Which term describes the smallest repeating unit that defines the crystal structure?**
    a. Unit cell
    b. Lattice vector
    c. Basis
    c. Crystal plane
2. **What is the purpose of Miller indices in crystallography?**
    a. To describe crystal symmetry elements
    b. To define the lattice translation vectors
    c. To classify crystal structures
    d. To specify crystallographic planes
3. **Which crystal structure is characterized by a cubic arrangement of atoms with chloride ions at the corners and sodium ions at the center of each face of the cube?**
    a. Diamond structure
    b. Hexagonal close-packed structure
    c. Sodium chloride structure
    d. Body-centered cubic structure
4. **What is the total number of symmetry operations (including identity) possible for a crystal?**
    a. 7
    b. 14
    c. 21
    d. 32
5. **Which symmetry operation involves a combination of rotation and inversion about a point?**
    a. Reflection
    b. Inversion
    c. Rotation axis
    d. Rotation-inversion axis
6. **What is the significance of symmetry in crystal structures?**
    a. It affects only the visual appearance of crystals.
    b. It determines mechanical properties but not optical properties.
    c. It influences physical properties such as conductivity and optical behavior.
    d It is irrelevant to the study of crystal growth.

Answers to Multiple Questions

1. (a); 2. (d); 3. (c); 4. (d); 5. (d); 6. (c)

REFERENCES

1. Acharya, K. R., Shapiro, R., Allen, S. C., Riordan, J. F., & Vallee, B. L. (1994). Crystal structure of human angiogenin reveals the structural basis for its functional divergence from ribonuclease. Proceedings of the National Academy of Sciences, 91(8), 2915–2919.
2. Bandyopadhyay, S., & Saha, S. (2007). GAPS: A clustering method using a new point symmetry-based distance measure. Pattern recognition, 40(12), 3430–3451.
3. Barlaud, M., Solé, P., Gaidon, T., Antonini, M., & Mathieu, P. (1994). Pyramidal lattice vector quantization for multiscale image coding. IEEE Transactions on Image Processing, 3(4), 367–381.
4. Barredo, D., Lienhard, V., De Leseleuc, S., Lahaye, T., & Browaeys, A. (2018). Synthetic three-dimensional atomic structures assembled atom by atom. Nature, 561(7721), 79–82.
5. Bates, F. S., Muthukumar, M., Wignall, G. D., & Fetters, L. J. (1988). Thermodynamics of isotopic polymer mixtures: Significance of local structural symmetry. The Journal of chemical physics, 89(1), 535–544.
6. Baur, W. H., & Tillmanns, E. (1986). How to avoid unnecessarily low symmetry in crystal structure determinations. Acta Crystallographica Section B: Structural Science, 42(1), 95–111.
7. Belch, A. C., Berkowitz, M., & McCammon, J. A. (1986). Solvation structure of a sodium chloride ion pair in water. Journal of the American Chemical Society, 108(8), 1755–1761.
8. Bilby, B. A. (1953). On the mutual transformation of lattices. The London, Edinburgh, and Dublin Philosophical Magazine and Journal of Science, 44(354), 782–785.
9. Brading, K., & Brown, H. R. (2004). Are gauge symmetry transformations observable?. The British journal for the philosophy of science, 55(4), 645–665.
10. Bragg, W. H., & Bragg, W. L. (1913). The structure of the diamond. Proceedings of the Royal Society of London. Series A, Containing Papers of a Mathematical and Physical Character, 89(610), 277–291.
11. Cantrell, J. H. (1994). Crystalline structure and symmetry dependence of acoustic nonlinearity parameters. Journal of applied physics, 76(6), 3372–3380.
12. Caulton, A. (2015). The role of symmetry in the interpretation of physical theories. Studies in History and Philosophy of Science Part B: Studies in History and Philosophy of Modern Physics, 52, 153–162.
13. Chakrabarti, A., Römer, R. A., & Schreiber, M. (2003). Magnetotransport in periodic and quasiperiodic arrays of mesoscopic rings. Physical Review B, 68(19), 195417.

14. Conway, J. H., & Sloane, N. J. (1992). Low-dimensional lattices. VI. Voronoi reduction of three-dimensional lattices. Proceedings of the Royal Society of London. Series A: Mathematical and Physical Sciences, 436(1896), 55–68.

15. Cowley, R. A. (1965). Raman scattering from crystals of the diamond structure. Journal de Physique, 26(11), 659–667.

16. Davi, M., Anne Thomas Doyle, M., & Proffitt, D. R. (1992). The role of symmetry in determining perceived centers within shapes. Perception & psychophysics, 52(2), 151–160.

17. De Wolff, P. M. (1974). The pseudo-symmetry of modulated crystal structures. Acta Crystallographica Section A: Crystal Physics, Diffraction, Theoretical and General Crystallography, 30(6), 777–785.

18. Domb, C. (1963). Excluded-volume effect for two-and three-dimensional lattice models. The Journal of Chemical Physics, 38(12), 2957–2963.

19. Eady, C., Lindsey, K., & Twell, D. (1995). The significance of microspore division and division symmetry for vegetative cell-specific transcription and generative cell differentiation. The Plant Cell, 7(1), 65–74.

20. Efremidis, N. K., Hudock, J., Christodoulides, D. N., Fleischer, J. W., Cohen, F. O., & Segev, M. (2003). Two-dimensional optical lattice solitons. Physical review letters, 91(21), 213906.

21. Ellis, R. G., & Oldenburg, D. W. (1994). Applied geophysical inversion. Geophysical Journal International, 116(1), 5–11.

22. Feng, Y., Sagaut, P., & Tao, W. (2015). A three dimensional lattice model for thermal compressible flow on standard lattices. Journal of Computational Physics, 303, 514–529.

23. Fisher, D., & Zimmer, R. J. (2002). Geometric lattice actions, entropy and fundamental groups. Commentarii Mathematici Helvetici, 77(2), 326–338.

24. Forrest, L. R. (2015). Structural symmetry in membrane proteins. Annual review of biophysics, 44(1), 311–337.

25. Gabrielse, G. (1980). Significance of time-reversal symmetry for time-resolved measurements of hydrogenic and other atomic observables. Physical Review A, 22(1), 138.

26. Gao, M. C., Zhang, B., Guo, S. M., Qiao, J. W., & Hawk, J. A. (2016). High-entropy alloys in hexagonal close-packed structure. Metallurgical and Materials Transactions A, 47, 3322–3332.

27. Gardner, R. L., & Davies, T. J. (2006). An investigation of the origin and significance of bilateral symmetry of the pronuclear zygote in the mouse. Human reproduction, 21(2), 492–502.

28. Gilbert, T. L. (1962). Inverse overlap matrix for periodic arrays of atoms. Journal of Mathematical Physics, 3(1), 107–117.

29. Halvorsen, K., Lesser, M., & Lundberg, A. (1999). A new method for estimating the axis of rotation and the center of rotation. Journal of biomechanics, 32(11), 1221–1227.

30. Harker, D. A. V. I. D. (1976). A table of the colored crystallographic and icosahedral point groups, including their chirality and diamorphism. Acta Crystallographica Section A: Crystal Physics, Diffraction, Theoretical and General Crystallography, 32(1), 133–139.
31. Hollister, A. M., Jatana, S., Singh, A. K., Sullivan, W. W., & Lupichuk, A. G. (1993). The axes of rotation of the knee. Clinical Orthopaedics and Related Research®, 290, 259–268.
32. Hu, P., Xu, P., Chen, B., & Wu, Q. (2017). A self-calibration method for the installation errors of rotation axes based on the asynchronous rotation of rotational inertial navigation systems. IEEE Transactions on Industrial Electronics, 65(4), 3550–3558.
33. Imry, Y., & Gunther, L. (1971). Fluctuations and physical properties of the two-dimensional crystal lattice. Physical Review B, 3(11), 3939.
34. Ismael, J., & Van Fraassen, B. C. (2003). Symmetry as a guide to superfluous theoretical structure. Symmetries in physics: Philosophical reflections, 371–392.
35. Janssen, T. (1969). Crystallographic groups in space and time: III. Four-dimensional Euclidean crystal classes corresponding to generalized magnetic point groups. Physica, 42(1), 71–92.
36. Jose, S., Surendran, P., Wang, Y., Herrera, I., Krzemien, L., Whitlock, S., ... & Hannaford, P. (2014). Periodic array of Bose-Einstein condensates in a magnetic lattice. Physical Review A, 89(5), 051602.
37. Kimball, G. E. (1935). The electronic structure of diamond. The Journal of Chemical Physics, 3(9), 560–564.
38. Kirsh, D. V., & Kupriyanov, A. V. (2015). Modeling and identification of centered crystal lattices in three-dimensional space. In Proceedings of Information Technology and Nanotechnology (ITNT-2015), CEUR Workshop Proceedings . 1490, 162–170.
39. Klewe, B., & Pedersen, B. (1974). The crystal structure of sodium chloride dihydrate. Acta Crystallographica Section B: Structural Crystallography and Crystal Chemistry, 30(10), 2363–2371.
40. Kojić-Prodić, B., & Štefanić, Z. (2010). Symmetry versus asymmetry in the molecules of life: homomeric protein assemblies. Symmetry, 2(2), 884–906.
41. Kopský, V. (1979). Tensorial covariants for the 32 crystal point groups. Acta Crystallographica Section A: Crystal Physics, Diffraction, Theoretical and General Crystallography, 35(1), 83–95.
42. Kovesi, P. (1997, December). Symmetry and asymmetry from local phase. In Tenth Australian joint conference on artificial intelligence . 190, 2–4.
43. Lei, L., Miao, L., Li, C., Liang, X., & Wang, J. (2021). The effects of composite primitive cells on band gap property of locally resonant phononic crystal. Modern Physics Letters B, 35(20), 2150334.
44. Lim, T., Mirabedini, P. S., Jung, K., Greaney, P. A., & Martinez-Morales, A. A. (2021). High-index crystal plane of ZnO nanopyramidal structures: Stabilization, growth, and improved photocatalytic performance. Applied Surface Science, 536, 147326.

45. Liu, Y., Liu, Y., & Drew, M. G. (2017). Comparison of calculations for interplanar distances in a crystal lattice. Crystallography Reviews, 23(4), 252–301.

46. Matusov, E., & Ney, H. (2010). Lattice-based ASR-MT interface for speech translation. IEEE transactions on audio, speech, and language processing, 19(4), 721–732.

47. Mondal, C., Agrawal, B. K., De, J. N., Samaddar, S. K., Centelles, M., & Viñas, X. (2017). Interdependence of different symmetry energy elements. Physical Review C, 96(2), 021302.

48. Mozrzymas, J. (1974). Crystal symmetry groups and their representations I lattice extensions of point groups in R2 and R3. Reports on Mathematical Physics, 6(1), 29–82.

49. Murgueitio Ramírez, S. (2022). A puzzle concerning local symmetries and their empirical significance. The British Journal for the Philosophy of Science, 73(4), 1021–1044.

50. Nakatsuka, A., Yoshiasa, A., Yamanaka, T., & Ohtaka, O. (1999). Symmetry change of majorite solid-solution in the system Mg3Al2Si3O12-MgSiO3. American Mineralogist, 84(7–8), 1135–1143.

51. Otte, H. M., & Crocker, A. G. (1965). Crystallographic formulae for hexagonal lattices. physica status solidi (b), 9(2), 441–450.

52. Page, Y. L. (1992). Ab-initio primitive cell parameters from single convergent-beam electron diffraction patterns: A converse route to the identification of microcrystals with electrons. Microscopy research and technique, 21(2), 158–165.

53. Pan, Z., Ma, R., Wang, D., & Chen, A. (2018). A review of lattice type model in fracture mechanics: theory, applications, and perspectives. Engineering Fracture Mechanics, 190, 382–409.

54. Partridge, P. G. (1967). The crystallography and deformation modes of hexagonal close-packed metals. Metallurgical reviews, 12(1), 169–194.

55. Preskill, J., Trivedi, S. P., Wilczek, F., & Wise, M. B. (1991). Cosmology and broken discrete symmetry. Nuclear Physics B, 363(1), 207–220.

56. Reading, N., Speyer, D. E., & Thomas, H. (2021). The fundamental theorem of finite semidistributive lattices. Selecta Mathematica, 27, 1–53.

57. Robins, A. J. (1991). Reflection of a plane wave from a fluid layer with continuously varying density and sound speed. The Journal of the Acoustical Society of America, 89(4), 1686–1696.

58. Savin, A., Jepsen, O., Flad, J., Andersen, O. K., Preuss, H., & von Schnering, H. G. (1992). Electron localization in solid-state structures of the elements: the diamond structure. Angewandte Chemie International Edition in English, 31(2), 187–188.

59. Smardzewski, J., & Wojciechowski, K. W. (2019). Response of wood-based sandwich beams with three-dimensional lattice core. Composite Structures, 216, 340–349.

60. Steed, J. W. (2003). Should solid-state molecular packing have to obey the rules of crystallographic symmetry?. CrystEngComm, 5(32), 169–179.

61. Steinberger, B., & O'Connell, R. J. (1997). Changes of the Earth's rotation axis owing to advection of mantle density heterogeneities. Nature, 387(6629), 169–173.

62. Tari, S. Z. (1999, July). Coloring 3D symmetry set: perceptual meaning and significance in 3D. In Visual Information Processing VIII . 3716 , 105–109).

63. Ter Horst, A. C., Van Lier, R., & Steenbergen, B. (2010). Mental rotation task of hands: differential influence number of rotational axes. Experimental Brain Research, 203, 347–354.

64. Tornari, V. (2022). A symmetry concept and significance of fringe patterns as a direct diagnostic tool in artwork conservation. Light: Advanced Manufacturing, 3(2), 236–257.

65. Tung, Y. T., Chen, L. J., Chen, Y. P., & Lin, S. T. (2012). Molecular dynamics study on the growth of structure I methane hydrate in aqueous solution of sodium chloride. The Journal of Physical Chemistry B, 116(48), 14115–14125.

66. Vaida, M., Shimon, L. J. W., Weisinger-Lewin, Y., Frolow, F., Lahav, M., Leiserowitz, L., & McMullan, R. K. (1988). The structure and symmetry of crystalline solid solutions: a general revision. Science, 241(4872), 1475–1479.

67. Wang, L., Wang, J., Gao, X., Chen, C., Da, Y., Wang, S., ... & Wu, Y. (2022). Periodic one-dimensional single-atom arrays. Journal of the American Chemical Society, 144(35), 15999–16005.

68. Wei, B., Ni, S., Liu, Y., & Song, M. (2019). Three dimensional crystallographic orientation relationships for hexagonal close packed structure to face centered cubic structure transformation in pure titanium. Scripta Materialia, 169, 46–51.

69. White, M., & Foulds, F. (2018). Symmetry is its own reward: on the character and significance of Acheulean handaxe symmetry in the Middle Pleistocene. antiquity, 92(362), 304–319.

70. Wilson, A. J. C. (1990). Space groups rare for organic structures. II. Analysis by arithmetic crystal class. Acta Crystallographica Section A: Foundations of Crystallography, 46(9), 742–754.

71. Yeh, J. W., Lin, S. J., Chin, T. S., Gan, J. Y., Chen, S. K., Shun, T. T., ... & Chou, S. Y. (2004). Formation of simple crystal structures in Cu-Co-Ni-Cr-Al-Fe-Ti-V alloys with multiprincipal metallic elements. Metallurgical and Materials Transactions A, 35, 2533–2536.

72. Zhang, G., Saeed, S., Aslam, A., Kanwal, S., Idrees, N., & Sumbal Shaikh, T. (2022). Study of HCP (Hexagonal Close-Packed) Crystal Structure Lattice through Topological Descriptors. Computational Intelligence and Neuroscience, 2022(1), 6069666.

CHAPTER 3

Chemical Bonding in Solids

LEARNING OBJECTIVES

After studying this chapter, you will be able to:

- Understand the principles of chemical bonding in solids;
- Understand the concept of ionic bonds as electron transfer between atoms;
- Explain the factors influencing ionic bond formation and stability;
- Explain the concept of covalent bonding as electron sharing between atoms;
- Understand how molecular orbital theory describes bonding and anti-bonding states;
- Understand metallic bonding and its characteristics in metals;
- Know the unique properties and importance of hydrogen bonds.

KEY TERMS FROM THIS CHAPTER

Bonding energy
Covalent bonds
Electronegativity
Ionic bonds
Molecular orbitals
Coulomb interaction
Delocalized electrons
Hydrogen bonds
Metallic bonding
Van der Waals forces

UNIT INTRODUCTION

The study of solid-state materials Physics explores the fascinating realm of matter where numerous atoms come together to form a tightly packed solid structure. This statement highlights the significant number of atoms involved, as it is on the order of . Upon initial observation, it may appear to be a daunting challenge to utilize conventional scientific approaches to describe the immense quantity of atoms involved. Nevertheless, the sheer abundance of atoms present enables us to quantitatively describe solids using innovative models designed specifically for this purpose. For these models to be successful, the atoms involved must not be selected randomly from the periodic table (Dronskowski & Blöchl, 1993). Instead, the solid must consist of a specific set of elements, with their atoms arranged in a structured manner in space. For solid-state physicists, the most important objects of study are the "elementary crystals," which are three-dimensional groupings of atoms of a particular kind or chemical compounds consisting of two elements. To comprehend solids and their distinctive features, it is necessary to initially attain a fundamental grasp of two specific occurrences: The first aspect pertains to the cohesive forces that bind the atoms of a solid together, specifically referring to the chemical bonding among atoms. The second crucial factor is the arrangement of atoms within the solid's structure (Pauling, 1981).

Chemists often consider several sorts of chemical bonds according to the kinds of atoms involved, particularly their position on the periodic table and their electronegativity, which refers to their ability to attract electrons. In this chapter, one will explore various types of bonds found in solids, including covalent bonds, van der Waals ionic bonds, links formed by fluctuating dipoles or atoms, hydrogen bonds, and metallic bonds. Indeed, these various aspects of the Schrodinger equation can be observed in different materials, each showcasing a combination of these bonding types. However, it is quite helpful to discuss these various types of bonds to gain a better understanding of the way chemical bonding can take place (Alemany & Canadell, 2014).

3.1. IONIC BONDS

In certain compounds, like binary compounds like NaCl, there is a concept called an ionic bond. It occurs when an electron travels from one atom to another, resulting in two ions with opposite charges that are attracted to each other due to their energy favorability. A chemical "reaction" of this form is typically written as:

$$Na + Cl \rightarrow Na^{+} + Cl^{-} \rightarrow NaCl.$$

To determine whether or not such a reaction occurs, it is necessary to examine the energy changes that occur during the transfer of electrons (Spruijt et al., 2012).

For a more detailed analysis of energetics, one can easily envision resolving the Schrodinger equation over an individual atom to determine the energy of the neutral atom, the positive ion, and the negative ion (Berkowitz et al., 2002). On the other hand, if finding a solution to the Schrodinger equation becomes:

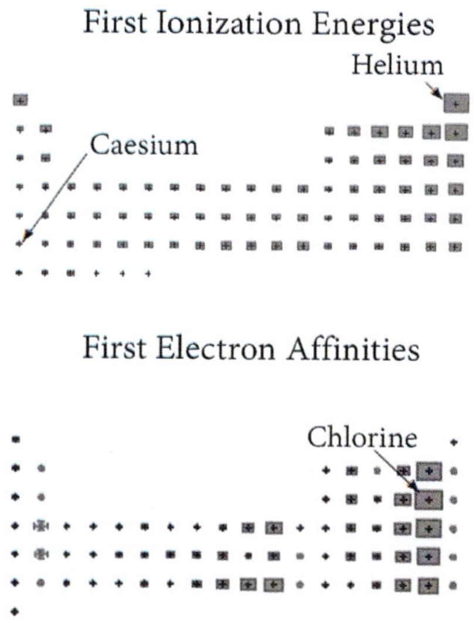

Figure 3.1. Illustration of pictorial periodic tables of first ionization energies (top) and first electron affinities (bottom).

Source: Steven H, Creative Commons License.

It can be quite challenging, as there are numerous methods available to measure the amount of energy of the atoms (or ions) using various spectroscopy techniques. Now define:

Specifically, in both scenarios, one is contrasting the energy associated with the electron being located at an infinite distance from the atom versus being located on the atom. In addition, if one is eliminating or adding just one electron, these are commonly referred to as first ionization energies and first electron affinities (Reed & Schleyer, 1990). Similarly, energies can be defined when electrons are removed or introduced, known as second ionization energies. Chemists commonly operate with systems maintained at a constant temperature and pressure, such as room temperature and atmospheric pressure. In such cases, their primary focus is usually on Gibbs free energies instead of pure energies. One will consistently presume that the individual is utilizing the suitable free energy source for the specific experiment at hand and will be imprecise and consistently refer to it as energy E (Taber et al., 2012). The ionization energy exhibits a periodic tendency, wherein it is minimal on the left side (groups I and II) from the periodic table or maximal on the right side (groups VII or VIII). The ionization energy likewise exhibits a reduction towards the lower regions of the periodic table, but to a smaller degree (see Figure 3.1). Similarly, the electron affinity is highest in the upper right corner of the periodic table, except for the noble gases in group VIII, which have very little measurable attraction for electrons (refer to Figure 3.1) (Saleh & Oganov, 2016).

The net energy change resulting from the transfer of an electron between atoms A onto atom B is:

$$\Delta E_{A+B\rightarrow A^{+}+B^{-}} = (Ionization\ Energy)_{A} - (Electron\ Affinity)_{B}.$$

Pay close attention to the sign. There is a positive energy required for ionization, while electron affinity releases energy. Nevertheless, this ΔE represents the energy required to move an electron from one atom to another that is located at a considerable distance from each other. Furthermore, there is also:

$$Cohesive\ Energy = Energy\ gain\ from\ A^{+} + B^{-} \rightarrow AB.$$

The cohesive energy primarily arises from the classical Coulomb interaction between the ions when they are brought into proximity (Luxford & Bretz, 2013). The overall energy gain for creating a molecule from two separate atoms is therefore determined by:

$$\Delta E_{A+B\rightarrow AB} = (Ionization\ Energy)_{A} - (Electron\ Affinity)_{B} - (Cohesive\ Energy\ of\ AB).$$

If the total ΔE over this process is lower than zero, an ionic bond is formed. When trying to figure out if an electron will move from one atom to another, one can use electronegativity as a handy tool. Electronegativity gives us an idea of how eager an atom is to have electrons or how strongly it pulls electrons towards itself. There are different ways to define electronegativity, but one commonly used definition refers to Mulliken Electronegativity. It is a simple and practical definition (Meot, 2005).

$$(Mulliken)\ Electronegativity = \frac{Electron\ Affinity + Ionization\ Energy}{2}.$$

Based on the regular patterns covered in Section 5.3, elements found in the upper right corner of the periodic table (excluding noble gases) exhibit a significantly high electronegativity (Wang et al., 2020). When it comes to bonding, electrons tend to move from atoms with lower electronegativity to those with higher electronegativity. The extent to which an electron is transferred from one atom to another is directly influenced by the disparity in electronegativity between the two atoms. When there is a small difference in electronegativity, the electron is partially transferred from a single atom to another. In the following section, it will be shown that a covalent bond can take place between two identical atoms, even in the absence of any difference in electronegativity. As a result, there is no exchange of electrons (Takagi, 1995). Before concluding the discussion on ionic bonding, it is worthwhile to delve into the characteristic physics shown by ionic solids. Initially, the materials exhibit a characteristic of being rigid and possess elevated melting points due to the robust Coulomb reaction among ions of opposite charges. Nevertheless, because of water's strong polarity, it can dissolve an ionic solid. This process (as shown in Figure 3.2) involves the organization of water molecules, where the negative end of each molecule is positioned close to the positive ions, while the positive end of the molecule is positioned close to the negative ions. In a solid with ionic bonding, the charges are strongly attached to the ions, leading to these substances having poor electrical conductivity (Zhu et al., 2018).

One way to understand this electronegativity is to consider it as the opposite of the chemical potential.

$$\frac{1}{2}\left(E_{affinity}+E_{ion}\right)=\frac{1}{2}\left(\left[E_N-E_{N+1}\right]+\left[E_{N-1}-E_N\right]\right)=\frac{\left[E_{N-1}-E_{N+1}\right]}{2}$$
$$\approx -\frac{\partial E}{\partial N}\approx -\mu.$$

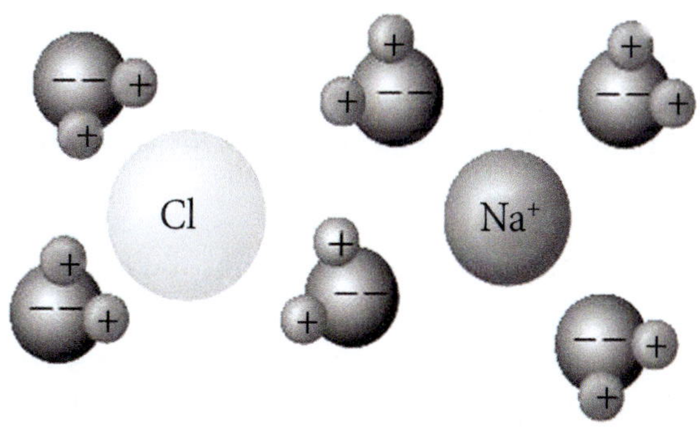

Figure 3.2. Illustration of salt, NaCl, dissolved in water.

Source: Bardeen, John, Creative Commons License.

3.2. COVALENT BOND

In a covalent bond, electrons are shared in a way that ensures a rough equality across two atoms. Various illustrations may be utilized to depict the covalent bond (Kamaras et al., 2003).

3.2.1. Particle in a Box Picture

Imagine a hydrogen atom as a box with a size of L for an electron (let's keep it simple and consider a one-dimensional system). The energy for a single electron within a box is quite recognizable (Sickles, 2009).

$$E = \frac{\hbar^2 \pi^2}{2mL^2}$$

Now imagine two atoms that come into proximity. An electron that is shared within two atoms can spread out over the relative positions within both atoms, resulting in a larger region for it to occupy and a decrease in its energy level (Mannila et al., 2001).

$$E = \frac{\hbar^2 \pi^2}{2m(2L)^2}$$

The decrease in energy produced through the electron's delocalization serves as the driving factor behind the formation of the bond. The most recent ground-state orbital is commonly referred to as a bonding orbital (Drake et al., 1988) (Figure 3.3).

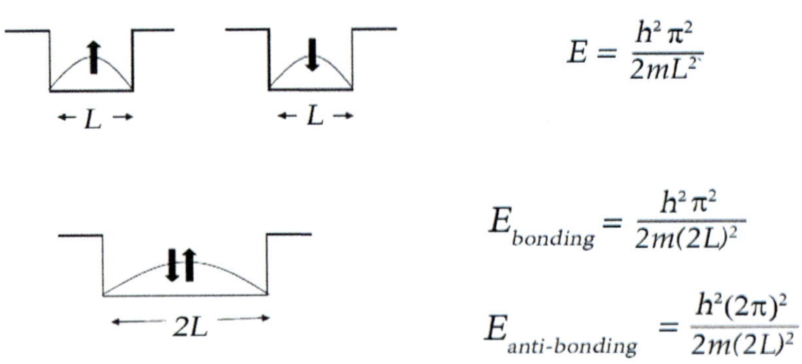

Figure 3.3. Illustration of Particle in a box picture of covalent bonding.

Source: Born, Max, Creative Commons License.

When two atoms combine to create a lower energy orbital, the electrons from each atom can occupy the same ground-state orbital, as they have opposite spin states. The method of bonding is illustrated through the molecular orbital picture displayed at the top of Figure 3.4. Naturally, the decrease in energy of both electrons needs to contend with the Coulomb repulsion that exists between two nuclei, as well as the Coulomb repulsion within the two electrons themselves. This calculation is considerably more intricate (Raduta et al., 2014).

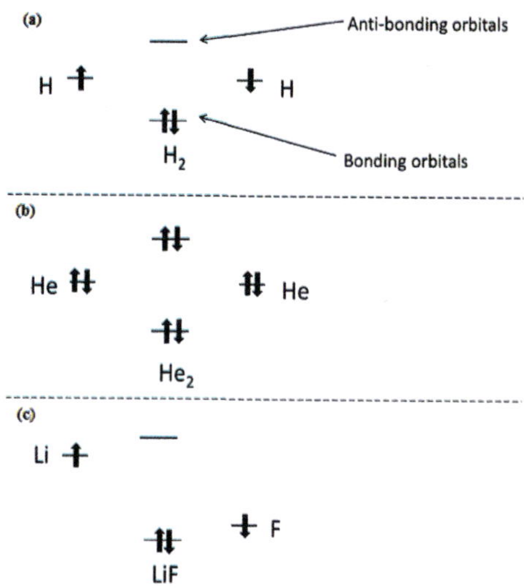

Figure 3.4. Illustration of molecular orbital picture of bonding. (a) Two hydrogen atoms (one having a spin-up electron and one having spin-down) come together to form a molecule (b) In the case of helium, since there are two electrons per atom, the bonding orbitals are filled, and the anti-bonding orbitals must be filled as well. The total energy is not reduced by the two helium atoms coming together (c) In the case of LiF, the energies of the lithium and the fluorine orbitals are different. As a result, the bonding orbital is mostly composed of the orbital on the F atom.

Source: William Lawrence, Creative Commons License.

Picture a situation where one start using two atoms of helium, each carrying two electrons. But when these two atoms come together, there is not enough room for just one ground-state wave function. In this case, two of the four electrons occupy the initial excited orbital, which coincidentally possesses the same electrical strength that of the ground-state orbital in the original atoms (Berbaum et al., 1983). When two atoms appear together, those excited orbitals are referred to as anti-bonding orbitals because no energy can be extracted by these electrons. (In reality, it takes energy to bring the two atoms closer together, taking into account the repulsion between their nuclei.) The molecular orbital illustration can be observed in the center of Figure 3.4 (Itoh & Odaka, 1991).

3.2.2. Molecular Orbital or Tight Binding Theory

In this section, one delves into the quantitative aspects of the concepts discussed earlier. Let's construct a Hamiltonian for a pair of hydrogen atoms. Given the disparity in weight between the nuclei and electrons, one can establish the fixed positions of the nuclei and proceed with solving the Schrodinger equation regarding the electrons, taking into account the varying distance between the nuclei. The fixation of the relative position of the nuclei is commonly referred to as the "Born-Oppenheimer" approximation. One aim is to determine the Eigen energies for the system by studying the relationship between the distances for the positively charged nuclei (Löwdin, 1962).

To simplify the discussion, one will focus on one electron and two similar positive nuclei. Now express the Hamiltonian in a specific form:

$$H = K + V_1 + V_2$$

Along with:

$$K = \frac{\mathbf{p}^2}{2m}$$

The electron's kinetic energy is an important factor in this context.

$$V_i = \frac{e^2}{4\pi\epsilon_0 |\mathbf{r} - \mathbf{R}_i|}$$

The Coulomb energy from interactions is determined by the distance between the electron or the nucleus .

In general, solving this particular Schrodinger equation exactly is challenging (while it is possible to do so in this instance, the solution does not provide significant insights). Instead, one will endeavor to find a variational solution. Let's express a trial wave function as:

$$|\psi\rangle = \phi_1|1\rangle + \phi_2|2\rangle \tag{3.1}$$

The set of coefficients along with the kets $|1\rangle$ and $|2\rangle$ are commonly referred to as "atomic" orbitals or "tight binding" orbitals. Equation 3.1 is often referred to as a "linear combination of atomic orbitals" or LCAO. The orbitals one is discussing can be considered as the solution to the Schrodinger equation with the existence of a single nucleus:

$$\begin{aligned}(K + V_1)|1\rangle &= \epsilon_0|1\rangle \\ (K + V_2)|2\rangle &= \epsilon_0|2\rangle\end{aligned} \tag{3.2}$$

Here, represents the energy for the single atom in its lowest state. In other words, $|1\rangle$ refers to the orbital in which an electron is bound to nucleus 1, while $|2\rangle$ represents the orbital where an electron is bound to nucleus 2. Let's assume that 2 and 1 are orthogonal for a rough approximation, allowing us to select a normalization (Akimov, 2016).

$$\langle i|j\rangle = \delta_{ij} \quad . \tag{3.3}$$

As the two nuclei approach each other closely, it becomes evident that this orthogonality is far from accurate. Fortunately, the majority that one learns is not heavily influenced by the orthogonality of orbitals. Therefore, for the sake of simplicity, one will assume that the orbitals are now orthonormal.

A highly efficient Schroedinger equation may be derived for the variational wave function, which, as expected, can be formulated as an Eigen value problem (Nishimoto et al., 2015).

$$\sum_j H_{ij}\phi_j = E\phi_i$$

Whereas:

$$H_{ij} = \langle i|H|j\rangle \tag{3.4}$$

In this case, one has a two-by-two matrix (the equation can be generalized to cases with over two orbitals). Remembering the concept of $|1\rangle$ referring to the ground-state energy for one may define it as:

$$H_{11} = \langle 1|H|1\rangle = \langle 1|K + V_1|1\rangle + \langle 1|V_2|1\rangle = \epsilon_0 + V_{cross} \tag{3.5}$$

$$H_{22} = \langle 2|H|2\rangle = \langle 2|K + V_2|2\rangle + \langle 2|V_1|2\rangle = \epsilon_0 + V_{cross} \tag{3.6}$$

$$H_{12} = \langle 1|H|2\rangle = \langle 1|K + V_2|2\rangle + \langle 1|V_1|2\rangle = 0 - t \tag{3.7}$$

$$H_{21} = \langle 2|H|1\rangle = \langle 2|K + V_2|1\rangle + \langle 2|V_1|1\rangle = 0 - t^* \ . \tag{3.8}$$

At the outset:

$$V_{cross} = \langle 1|V_2|1\rangle = \langle 2|V_1|2\rangle$$

The Coulomb potential is experienced by orbital $|1\rangle$ as a result of nucleus 2, or by orbital 2 as a result of nucleus 1. In the following pair of lines (Eqs. 6.7 along 6.8), one has introduced what is known as the hopping term (Elstner & Seifert, 2014).

$$t = -\langle 1|V_2|2\rangle = -\langle 1|V_1|2\rangle.$$

The explanation for the term "hopping" will become apparent shortly. It is worth noting that for the next two lines (Eqs. 6.7 or 6.8), the first term disappears due to the orthogonality of $|1\rangle$ and $|2\rangle$ (also using Eqs. 6.2). Now, one has simplified our Schroedinger equation to a compact two-by-two matrix equation (Sutton et al., 1988).

$$\begin{pmatrix} \epsilon_0 + V_{cross} & -t \\ -t^* & \epsilon_0 + V_{cross} \end{pmatrix} \begin{pmatrix} \phi_1 \\ \phi_2 \end{pmatrix} = E \begin{pmatrix} \phi_1 \\ \phi_2 \end{pmatrix}. \tag{3.9}$$

This equation can be interpreted as follows: orbitals $|1\rangle$ and $|2\rangle$ have energies ϵ0 that are altered by because of the influence of the neighboring nucleus. Furthermore, the electron can move from one orbital to another through the off-diagonal t term. To have a deeper understanding of this interpretation, it becomes apparent that within the time-dependent Schrödinger equation, assuming the matrix was to possess diagonal elements, a wave function that initially occupies orbital 1 would be confined to that orbital indefinitely. Nevertheless, due to the presence of the off-diagonal factor, the time-dependent waveform is capable of oscillating between the two orbitals (Gaus et al., 2014).

By diagonalizing this 2x2 matrix, one can determine the Eigen energies:

$$E_{\pm} = \epsilon_0 + V_{cross} \pm |t|$$

The bonding orbital represents the lower energy orbital, while the anti-bonding orbital is the higher energy orbital. The wave functions that relate to this are then determined (Weaire & Thorpe, 1971).

$$\left|\psi_{bonding}\right\rangle = \frac{1}{\sqrt{2}}\left(|1\rangle \pm |2\rangle\right) \qquad (3.10)$$

$$\left|\psi_{antibonding}\right\rangle = \frac{1}{\sqrt{2}}\left(|1\rangle \mp |2\rangle\right). \qquad (3.11)$$

That is, these represent the balanced and unbalanced combinations of orbitals. The signs ± and ∓ are determined by the sign of t. The lower energy orbital is commonly referred to as the bonding orbital, while the higher energy orbital is known as the anti-bonding orbital. When $t > 0$, the lowest energy bonding orbital seems. One can roughly consider these two wave functions to be the lowest pair of orbitals in a "particle-in-a-box" system. This lowest-energy wave function remains constant in sign regardless of position, while the first excited state alters sign once, indicating the presence of a single node (Nikiforov et al., 2014). This analogy holds when $t > 0$.

Let's take a moment to explore what happens when two nuclei that are not identical become bound together. In this scenario, the energy over an electron to occupy orbital 1 would vary compared to orbital 2 (refer to the lower part of Figure 3.4). In the matrix's equation, Equation 3.9, the entries together on the diagonal would no longer be equal. Additionally, in the ground state, the magnitudes of and might no longer be equal as stated in Equation 3.10. On the other hand, the lower-energy orbital will be more heavily occupied in the ground state (Goedecker & Teter, 1995).

It is important to remember that represents the energy experienced by the electron in orbital 1 due to the presence of nucleus 2. However, it is important to note that the interaction across the two nuclei is also a significant factor. In a simplified model, the repulsion within the two nuclei counteracts the attractive energy within the nucleus or the electrons in the opposite orbital (Kambe, 1981). Therefore, by considering this energy, one will obtain:

$$\tilde{E}_\pm \approx \epsilon_0 \pm |t|.$$

As the nuclei approach each other, the hop term t rises, resulting in a level of energy diagram depicted in Figure 3.5. This image is implausible, as it implies that two atoms would form a bond at a distance of zero between their nuclei. The issue at hand arises when our presumptions and estimations start to deteriorate as the nuclei approach each other in proximity (such as the loss of orthogonality in our orbitals and the incomplete cancellation of Coulomb energy among nuclei by , among other factors) (Seifert, 2007).

Figure 3.6 presents a more accurate energy level diagram depicting the bonding along with anti-bonding states. It is important to observe that the energy increases without limit as the nuclei are brought closer together due to the repulsive force between them known as Coulomb repulsion. As a result, the system has a minimum energy when the nuclei are located at a distance greater than zero from each other. Subsequently, this gap transforms into the equilibrium distance across the nuclei of the resulting molecule (Lu et al., 2019).

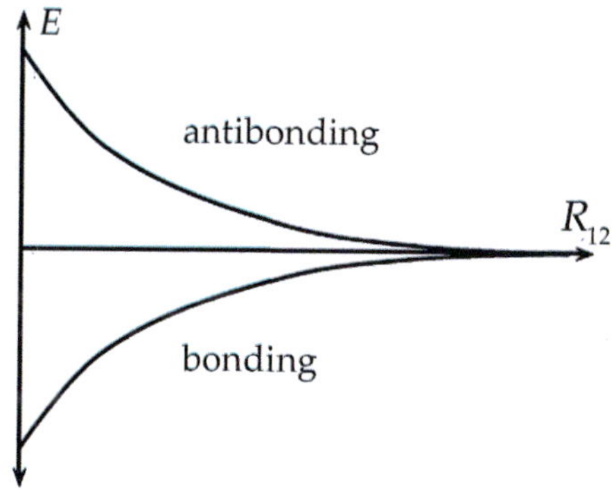

Figure 3.5. Illustration of model tight binding energy levels as a function of distance between the nuclei of the atoms.

Source: Landauer, Creative Commons License.

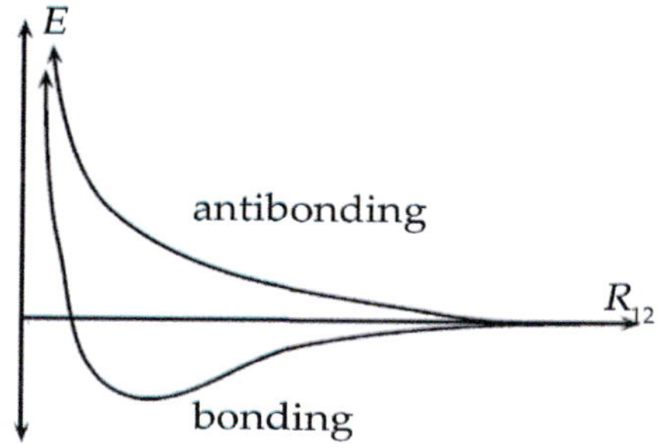

Figure 3.6. Illustration of more realistic energy levels as a function of distance between the nuclei of the atoms.

Source: Slater, John, Creative Commons License.

Figure 3.6 shows that there is a specific spacing between the nuclei where the bonding energy reaches its minimal value. The optimum distance refers to the link length between two atoms. Nevertheless, when the temperature is finite, the distance will fluctuate about the minimum value (similar to the behavior of a particle within a potential well with a non-zero temperature). Due to the asymmetry of the potential well, the "particle" confined within it will exhibit more fluctuations towards bigger distances compared to smaller distances at finite temperatures. Consequently, the average distance between bonds will rise when the temperature is not at absolute zero (Cui & Elstner, 2014).

Materials that are bonded through covalent bonds are known for their strength and their ability to function as electrical semiconductors or insulators. The orientation of the orbitals gives these materials a strong structural integrity, resulting in a lack of malleability and brittleness. They do not readily dissolve within polar solvents like water, unlike ionic materials (Ordejón, 1998).

3.3. VAN DER WAALS, FLUCTUATING DIPOLE FORCES, OR MOLECULAR BONDING

When two atoms (or two molecules) are separated by a significant distance, there is still an attractive force between them called the van der Waals force, also referred to as fluctuating dipole forces or molecular bonding. Simply put, both atoms possess a dipole moment that, while it may average out to zero, can experience temporary fluctuations as a result of quantum mechanics. When one atom acquires a temporary dipole moment, it can induce polarization in the second atom, causing it to develop its dipole moment to reduce its energy. Consequently, both atoms shall be drawn towards each other (Ashcroft, 1991).

The bonding between atoms of this nature is commonly observed in inert atoms, specifically noble gases like helium, neon, krypton, argon, and xenon. These atoms possess electrons that do not participate in either covalent or ionic bonding. Bonding within inert molecules, such as nitrogen molecules , is common. In these cases, the electrons within the molecule are unable to form either covalent or ionic interactions with other molecules. This type of bonding is relatively weaker when compared to bonds covalent or ionic. However, it is characterized by its long-range nature, as the electrons do not have to hop between atoms (Israelachvili, 1974).

For a more quantitative approach, let's examine the behavior of an electron as it revolves around a nucleus, such as a proton. When the electron remains in a fixed position, a dipole moment is formed, with r representing the vector that links the electron towards the proton.

When the electron is in the unperturbed eigenstate, its median dipole moment is zero. Nevertheless, when an electric charge is exerted on the atom, polarization occurs. This means that the electron is more likely to be located on one side of the nucleus rather than the other. One writes

$$p = \aleph E$$

where; χ is referred to as the polarization (also called electric susceptibility). One can explicitly calculate the polarizability of a hydrogen atom, for example. Regardless, it is a definite and non-negative value (Ambrosetti et al., 2016).

Now, imagine one has two atoms that are a certain distance apart in the x-direction (see Figure 3.7). Let's consider a scenario where an atom experiences a temporary dipole moment, denoted as p1, which we will assume is aligned in the -direction for clarity. Subsequently, the second atom is going to encounter an electric field (Delannay et al., 2007).

$$E = \frac{p_1}{4\pi\epsilon_0 r^3}$$

Heading towards the negative z-axis direction. As a result of its polarizability, a second atom creates a dipole moment with = χE, that is then attracted towards the first atom. The potential energy within these two dipoles can be calculated (Winterton, 1970).

$$U = \frac{-|p_1||p_2|}{4\pi\epsilon_0 r^3} = \frac{-p_1\chi E}{(4\pi\epsilon_0)r^3} = \frac{-|p_1|^2\chi}{(4\pi\epsilon_0 r^3)^2}. \quad (3.12)$$

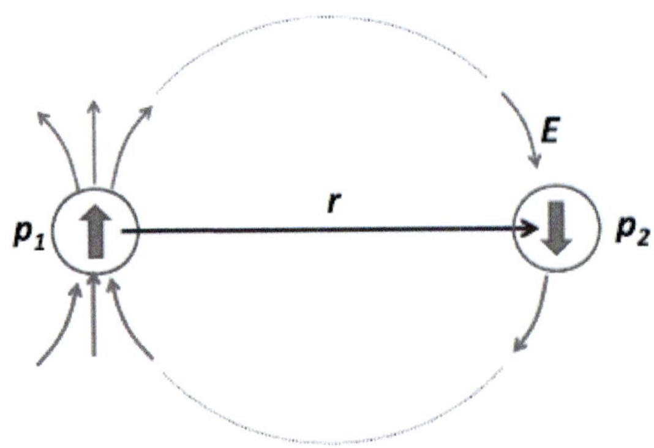

Figure 3.7. Illustration of an atom with a polarization induces a polarization in a second atom.

Source: Eugene, Creative Commons License.

Thus, there exists an attractive force that is directly proportional to .

It is worth noting that this argument seems to rely on the assumption how the dipole moment, p_1 for the very first atom is not zero, even though, on average, the dipole moment in the atom is actually zero. However, in equation 3.12, what actually comes into play is the squared magnitude of , that has a non-zero average value (Weiss et al., 2001).

However, these varying dipolar forces are typically not very strong and are the sole forces that arise when electrons are not shared or sent between atoms. This occurs either when the electrons remain chemically active or if the atoms are widely separated. When multiple atoms are combined, the collective force of van der Waals can become significantly stronger, despite their weakness. One fascinating instance of this van der Waals force has become the remarkable ability of certain creatures, like geckos, to effortlessly scale even the most polished surfaces, like glass windows. Their feet are covered in hair that allows for intimate contact via the atoms within the wall, resulting in their strong adhesion primarily because of van der Waals forces (Cole et al., 2009).

3.4. METALLIC BONDING

Differentiating between metallic bonding and covalent bonding can be challenging at times. Generally, a metallic bond is defined as the type of bonding that takes place within a metal. These bonds exhibit similarities to covalent bonds since they involve the sharing of electrons between atoms. However, in this scenario, the electrons disperse and get delocalized across the crystal structure. The delocalized electrons that are free might be seen as the cohesive force that binds the positive ions have left (Soler et al., 2000).

Due to the complete delocalization of electrons, the bonds within metals typically lack directionality. Metals are frequently characterized by their ductility and malleability. Metals possess excellent electrical and thermal conductivity due to the presence of freely moving electrons (Acar & Tarhan, 2008).

Did you know?

Metallic bonding facilitates excellent electrical conductivity in metals due to the presence of delocalized electrons that can move unrestrictedly throughout the substance. The distinct characteristic of metals is the reason why it gets indispensable for electrical wiring and numerous

3.5. HYDROGEN BONDS

The hydrogen atom seems exceptionally unique because of its minuscule dimensions. Consequently, the bonds established with hydrogen atoms exhibit distinct qualitative characteristics compared to other bonds. When a hydrogen atom makes a covalent and ionic connection with a larger atom, the tiny hydrogen nucleus, remains on the surface attached to the partner atom. Consequently, the molecule (composed of hydrogen and its counterpart) becomes a dipole. These dipoles can attract charges or other dipoles in a typical manner (Perrin & Nielson, 1997).

Hydrogen has a unique characteristic where, when it establishes a connection, its electron is drawn towards its partner, leaving behind a proton with a positive charge that is not shielded by electrons that are in the core orbitals. Consequently, this positive charge exhibits a high level of effectiveness in its attraction towards additional aggregations of electrons (Herschlag & Pinney, 2018). O, serves as an excellent illustration of the hydrogen bond.

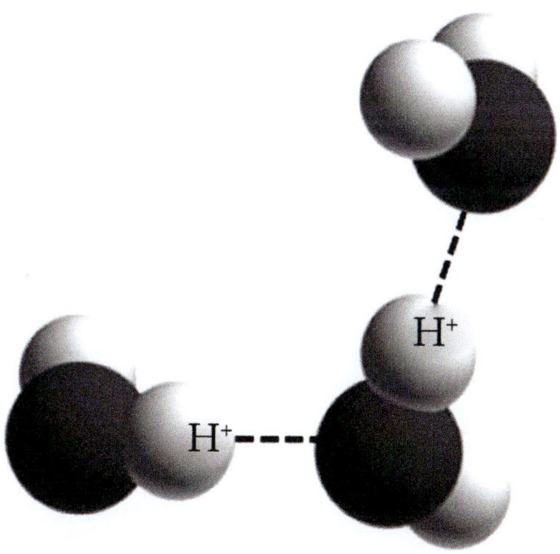

Figure 3.8. Illustration of hydrogen bonds in O.

Source: Simply, Creative Commons License) (Source: Sommerfeld, Creative Commons License.

The oxygen atoms are connected to two hydrogen atoms, although their arrangement is not in a straight line due to the atomic orbital structure. The positively charged hydrogen atoms are attracted to the oxygen atoms of neighboring water molecules. In ice, the attractions within water molecules are sufficiently strong to create a stable bond, resulting in the formation of a crystal structure. At times, it is possible to consider the hydrogen atom to be creating a connection between two oxygen atoms, effectively keeping them united (Alkorta et al., 1998). Hydrogen bonding plays a crucial role in biological molecules, such as the binding of DNA strands. The attraction between a hydrogen atom on a single water molecule and an oxygen atom on another molecule results in the formation of a relatively weak hydrogen bond. These bonds exhibit sufficient strength to allow for the formation of ice at temperatures below 273.15 K under standard atmospheric pressure (Steiner, 2002) (Figure 3.8).

Remember

It is important to note that hydrogen bonds, while less strong than covalent or ionic connections, play a critical role in the formation of biological molecules such as DNA and proteins. Which enhances the stability and functionality of these molecules, underscoring their significance in biological systems.

ACTIVITY 3.1

Objective: To demonstrate and differentiate between ionic and covalent bonding using simple materials.

Instructions:

1. Use Red and Blue Modeling Clay to Create Two Atoms of Each Color.
   - **For Ionic Bonds:** Connect one red and one blue atom using toothpicks to represent electron transfer and ion formation.
   - For Covalent Bonds: Connect two atoms of the same color with toothpicks to demonstrate electron sharing and bond formation.
2. Discuss the Differences Between Ionic and Covalent Bonds:
   - **Ionic Bonds:** Explain how electrons are transferred, forming ions that attract each other due to opposite charges.
   - **Covalent Bonds:** Highlight the sharing of electron pairs between atoms, emphasizing different bond strengths (single, double, triple).

SUMMARY

- The chapter on Chemical Bonding in Solids explores the fundamental principles behind different types of bonds and interactions that govern the behavior and properties of solid materials. It covers Ionic Bonds, where electrons are transferred between atoms to form ions that attract each other through electrostatic forces. Covalent Bonds are discussed for their role in sharing electron pairs between atoms, creating strong, directional bonds essential for the structural integrity of solids.
- The chapter also delves into theoretical frameworks such as Particle in a Box Picture and Molecular Orbital or Tight Binding Theory. Additionally, it examines Van der Waals and Fluctuating Dipole Forces.
- Additionally, Metallic Bonding is examined for its role in metals, where delocalized electrons move freely among positively charged metal ions, imparting unique electrical and thermal conductivity properties. Finally, Hydrogen Bonds are highlighted for their specific type of dipole-dipole interaction that plays a crucial role in stabilizing molecular structures in solids.

REVIEW QUESTIONS

1. Describe ionic bonds and provide an example of a compound formed by this type of bonding.
2. Explain covalent bonds and their importance in stabilizing solid-state materials. Give an example of a substance with predominantly covalent bonding.
3. How does the Particle in a Box Picture aid in understanding the electronic properties of solids? What does it reveal about electron behavior in confined spaces?
4. Compare Molecular Orbital Theory and Tight Binding Theory in explaining electronic structures of solids. Highlight their differences.
5. Discuss the role of Van der Waals forces in influencing solid properties. Give examples of materials where these forces are significant.
6. Differentiate metallic bonding from other types. Explain how electron delocalization contributes to unique properties of metals.
7. Explore the significance of hydrogen bonds in solid structures. Provide examples of materials where hydrogen bonding is crucial.

MULTIPLE CHOICE QUESTIONS

1. **What type of bonding is characterized by the transfer of electrons from one atom to another?**
   a. Covalent Bond
   b. Metallic Bond

c. Ionic Bond

d. Van der Waals Bond

2. **Which theoretical framework describes the electronic structure of solids based on atomic orbitals?**

a. Tight Binding Theory

b. Molecular Orbital Theory

c. Metallic Bonding Theory

d. Van der Waals Theory

3. **What type of bond involves the sharing of electron pairs between atoms?**

a. Ionic Bond

b. Covalent Bond

c. Metallic Bond

d. Hydrogen Bond

4. **Which bonding type is characterized by delocalized electrons moving freely throughout the metal structure?**

a. Ionic Bond

b. Covalent Bond

c. Metallic Bond

d. Van der Waals Bond

5. **What type of intermolecular force is responsible for temporary dipoles between atoms or molecules?**

a. Covalent Bond

b. Ionic Bond

c. Van der Waals Forces

d. Metallic Bond

Answers to Multiple Questions

1. (c); 2. (a); 3. (b); 4. (c); 5. (c)

REFERENCES

1. Acar, B., & Tarhan, L. (2008). Effects of cooperative learning on students' understanding of metallic bonding. Research in Science Education, 38, 401–420.

2. Akimov, A. V. (2016). Nonadiabatic molecular dynamics with tight-binding fragment molecular orbitals. Journal of Chemical Theory and Computation, 12(12), 5719–5736.

3. Alemany, P., & Canadell, E. (2014). Chemical bonding in solids. The Chemical Bond: Fundamental Aspects of Chemical Bonding, 445–476.

4. Alkorta, I., Rozas, I., & Elguero, J. (1998). Non-conventional hydrogen bonds. Chemical Society Reviews, 27(2), 163–170.

5. Ambrosetti, A., Ferri, N., DiStasio Jr, R. A., & Tkatchenko, A. (2016). Wavelike charge density fluctuations and van der Waals interactions at the nanoscale. Science, 351(6278), 1171–1176.

6. Ashcroft, N. W. (1991). Electronic fluctuation and the van der Waals metal. Philosophical Transactions of the Royal Society of London. Series A: Physical and Engineering Sciences, 334(1635), 407–423.

7. Berbaum, K., Tharp, D., & Mroczek, K. (1983). Depth perception of surfaces in pictures: Looking for conventions of depiction in Pandora's box. Perception, 12(1), 5–20.

8. Berkowitz, B., Klafter, J., Metzler, R., & Scher, H. (2002). Physical pictures of transport in heterogeneous media: Advection-dispersion, random-walk, and fractional derivative formulations. Water Resources Research, 38(10), 9–1.

9. Cole, M. W., Velegol, D., Kim, H. Y., & Lucas, A. A. (2009). Nanoscale van der Waals interactions. Molecular Simulation, 35(10–11), 849–866.

10. Cui, Q., & Elstner, M. (2014). Density functional tight binding: values of semi-empirical methods in an ab initio era. Physical Chemistry Chemical Physics, 16(28), 14368–14377.

11. Delannay, R., Louge, M., Richard, P., Taberlet, N., & Valance, A. (2007). Towards a theoretical picture of dense granular flows down inclines. Nature Materials, 6(2), 99–108.

12. Drake, T. G., Shreve, R. L., Dietrich, W. E., Whiting, P. J., & Leopold, L. B. (1988). Bedload transport of fine gravel observed by motion-picture photography. Journal of Fluid Mechanics, 192, 193–217.

13. Dronskowski, R., & Blöchl, P. E. (1993). Crystal orbital Hamilton populations (COHP): energy-resolved visualization of chemical bonding in solids based on density-functional calculations. The Journal of Physical Chemistry, 97(33), 8617–8624.

14. Elstner, M., & Seifert, G. (2014). Density functional tight binding. Philosophical Transactions of the Royal Society A: Mathematical, Physical and Engineering Sciences, 372(2011), 20120483.

15. Gaus, M., Cui, Q., & Elstner, M. (2014). Density functional tight binding: application to organic and biological molecules. Wiley Interdisciplinary Reviews: Computational Molecular Science, 4(1), 49–61.

16. Goedecker, S., & Teter, M. (1995). Tight-binding electronic-structure calculations and tight-binding molecular dynamics with localized orbitals. Physical Review B, 51(15), 9455.

17. Herschlag, D., & Pinney, M. M. (2018). Hydrogen bonds: simple after all?. Biochemistry, 57(24), 3338–3352.

18. Israelachvili, J. N. (1974). The nature of van der Waals forces. Contemporary Physics, 15(2), 159–178.

19. Itoh, T., & Odaka, K. (1991). A Particle-Picture Approach to Anomalies in Chiral Gauge Theories. Fortschritte der Physik/Progress of Physics, 39(8), 557–620.

20. Kamaras, K., Itkis, M. E., Hu, H., Zhao, B., & Haddon, R. C. (2003). Covalent bond formation to a carbon nanotube metal. Science, 301(5639), 1501–1501.

21. Kambe, K. (1981). A simple tight-binding theory with spin-orbit coupling for the analysis of two-dimensional band structures of adsorbates. Surface Science, 105(1), 95–113.

22. Löwdin, P. O. (1962). Band theory, valence bond, and tight-binding calculations. Journal of Applied Physics, 33(1), 251–280.

23. Lu, X., Duchimaza-Heredia, J., & Cui, Q. (2019). Analysis of density functional tight binding with natural bonding orbitals. The Journal of Physical Chemistry A, 123(34), 7439–7453.

24. Luxford, C. J., & Bretz, S. L. (2013). Moving beyond definitions: what student-generated models reveal about their understanding of covalent bonding and ionic bonding. Chemistry Education Research and Practice, 14(2), 214–222.

25. Mannila, K., Koponen, I. T., & Niskanen, J. A. (2001). Building a picture of students' conceptions of wave-and particle-like properties of quantum entities. European journal of physics, 23(1), 45.

26. Meot-Ner, M. (2005). The ionic hydrogen bond. Chemical reviews, 105(1), 213–284.

27. Nikiforov, I., Dontsova, E., James, R. D., & Dumitric⊠, T. (2014). Tight-binding theory of graphene bending. Physical Review B, 89(15), 155437.

28. Nishimoto, Y., Fedorov, D. G., & Irle, S. (2015). Third-order density-functional tight-binding combined with the fragment molecular orbital method. Chemical Physics Letters, 636, 90–96.

29. Ordejón, P. (1998). Order-N tight-binding methods for electronic-structure and molecular dynamics. Computational materials science, 12(3), 157–191.

30. Pauling, L. C. (1981). Early work on chemical bonding in relation to solid-state physics. Proceedings of the Royal Society of London. A. Mathematical and Physical Sciences, 378(1773), 207–218.

31. Perrin, C. L., & Nielson, J. B. (1997). "Strong" hydrogen bonds in chemistry and biology. Annual review of physical chemistry, 48(1), 511–544.

32. Raduta, A. A., Raduta, C. M., & Faessler, A. (2014). A new picture for the chiral symmetry properties within a particle–core framework. Journal of Physics G: Nuclear and Particle Physics, 41(3), 035105.

33. Reed, A. E., & Schleyer, P. V. R. (1990). Chemical bonding in hypervalent molecules. The dominance of ionic bonding and negative hyperconjugation over d-orbital participation. Journal of the American Chemical Society, 112(4), 1434–1445.

34. Saleh, G., & Oganov, A. R. (2016). Alkali subhalides: high-pressure stability and interplay between metallic and ionic bonds. Physical Chemistry Chemical Physics, 18(4), 2840–2849.

35. Seifert, G. (2007). Tight-binding density functional theory: an approximate Kohn–Sham DFT scheme. The Journal of Physical Chemistry A, 111(26), 5609–5613.

36. Sickles, A. (2009). A four component picture for jet induced correlations in heavy ion collisions? Recent results on two and three particle correlations. The European Physical Journal C, 61, 583–588.

37. Soler, J. M., Beltrán, M. R., Michaelian, K., Garzón, I. L., Ordejón, P., Sánchez-Portal, D., & Artacho, E. (2000). Metallic bonding and cluster structure. Physical Review B, 61(8), 5771.

38. Spruijt, E., van den Berg, S. A., Cohen Stuart, M. A., & van der Gucht, J. (2012). Direct measurement of the strength of single ionic bonds between hydrated charges. Acs Nano, 6(6), 5297–5303.

39. Steiner, T. (2002). The hydrogen bond in the solid state. Angewandte Chemie International Edition, 41(1), 48–76.

40. Sutton, A. P., Finnis, M. W., Pettifor, D. G., & Ohta, Y. (1988). The tight-binding bond model. Journal of Physics C: Solid-state physics, 21(1), 35.

41. Taber, K. S., Tsaparlis, G., & Nakibo⊠lu, C. (2012). Student conceptions of ionic bonding: Patterns of thinking across three European contexts. International Journal of Science Education, 34(18), 2843–2873.

42. Takagi, F. (1995). Consistency conditions for relativistic composite systems in a classical particle picture. Progress of Theoretical Physics, 94(3), 385–397.

43. Wang, X., Liang, D., & Cheng, B. (2020). Preparation and research of intrinsic self-healing elastomers based on hydrogen and ionic bond. Composites Science and Technology, 193, 108127.

44. Weaire, D., & Thorpe, M. F. (1971). Electronic properties of an amorphous solid. I. A simple tight-binding theory. Physical Review B, 4(8), 2508.

45. Weiss, M. S., Brandl, M., Sühnel, J., Pal, D., & Hilgenfeld, R. (2001). More hydrogen bonds for the (structural) biologist. Trends in biochemical sciences, 26(9), 521–523.

46. Winterton, R. H. S. (1970). Van der Waals forces. Contemporary Physics, 11(6), 559–574.

47. Zhu, M., He, S., Dai, Y., Han, J., Gan, L., Liu, J., & Long, M. (2018). Long-lasting sustainable self-healing ion gel with triple-network by trigger-free dynamic hydrogen bonds and ion bonds. ACS Sustainable Chemistry & Engineering, 6(12), 17087–17098.

CHAPTER 4

Phonons and Thermal Properties of Solids

LEARNING OBJECTIVES

After studying this chapter, you will be able to:

- Understand the concept of lattice heat capacity in crystals;
- Describe the Debye model and its application to phonon heat capacity;
- Explain the significance of density of states in phonon physics;
- Analyze the impact of anharmonic interactions on lattice vibrations in crystals;
- Explain thermal expansion using anharmonic potentials in solid materials;
- Explain the role of phonon mean free path in determining thermal conductivity;
- Describe umklapp processes and their impact on thermal resistivity in crystals.

KEY TERMS FROM THIS CHAPTER

Anharmonic interactions
Density of modes ($D(\omega)$)
Heat capacity and phonons:
Planck distribution function
Thermal equilibrium
Debye model
Density of states (DOS)
Lattice heat capacity
Thermal conductivity
Wavevectors

UNIT INTRODUCTION

For a wide range of real-world applications, from creating effective electrical devices to perfecting materials for thermal management systems, it is essential to comprehend how solids react to temperature fluctuations. Phonons, which are quantized vibrations inside the atomic structure of crystalline solids, are fundamental to these features. Important thermal properties including heat capacity, thermal conductivity, and thermal expansion are largely determined by phonons. In order to understand how materials store, transport, and respond to thermal energy, scientists and engineers study the behavior and interactions of phonons within the crystal lattice (Sanati et al., 2011).

In condensed matter physics and materials science, phonons and the thermal characteristics of solids constitute a fundamental field of study. Like the quantized vibrations of molecules in gases, phonons are collective excitations of atoms within a crystalline lattice. These vibrations move through the lattice and are essential for figuring out a solid's different thermal characteristics (Chiuri & Jappelli, 2003).

Thermal conductivity, or a material's ability to conduct heat efficiently, is one important factor. By moving vibrations from hot to cool spots inside the lattice, phonons move heat energy. Higher thermal conductivities are typically found in materials with longer mean free pathways between collisions and higher phonon frequencies. This characteristic is critical in applications such as electronics, where effective heat dissipation is critical, and thermoelectric materials, where the preservation of a thermal gradient is necessary for the conversion of heat into electricity (Nipko & Loong, 1998).

Furthermore, phonons have an impact on solids' specific heat capacity, which measures the quantity of heat needed to raise a material's temperature. Specific heat is contributed by several phonon vibrational modes, with low-frequency acoustic phonons usually predominating at low temperatures. Comprehending these contributions facilitates the forecasting and enhancement of thermal response under many circumstances, including high-temperature settings and cryogenic applications (Wang et al., 2016).

In addition, phonons have an impact on thermal expansion, which refers to the changes in a material's dimensions as a result of temperature variations. As the temperature rises, the magnitude of atomic vibrations (phonons) also increases, resulting in expansion. These phenomena is crucial in engineering materials for applications that require consistent dimensions even when exposed to temperature variations, such as in aeronautical constructions or precision equipment (Chen et al., 2018).

The chapter examines the basic ideas behind the phonons, or lattice vibrations that occur within solids. When it comes to figuring out a solid's heat capacity, thermal conductivity, and thermal expansion, phonons are essential. The theoretical models for phonon behavior and density of states, such as the Debye and Einstein models, are examined in this chapter. It also studies the effects of defects, scattering mechanisms, and anharmonic interactions on thermal phenomena and phonon transport in various crystal formations. Comprehending these ideas is fundamental to understanding how solids behave thermally at various temperatures and material compositions (Chernatynskiy et al., 2011).

4.1. PHONON HEAT CAPACITY

Since it is determined by experimentation, the heat capacity at constant volume is frequently seen as being more fundamental than the heat capacity at constant pressure. The heat capacity at constant volume is commonly defined as the derivative of internal energy with respect to temperature, while keeping the volume constant, denoted as . This definition allows us to understand how the energy of a system changes with temperature, while keeping the volume constant (Vali, 2008). The lattice heat capacity, represented by the symbol , is the thermal capacity of a crystal induced by phonons. The sum of the energies of all phonon modes, which are identified by wavevector K and polarization index p, yields the total energy of the phonons in a crystal at temperature τ, which is equivalent to .

$$U_{\text{lat}} = \sum_K \sum_p U_{K,p} = \sum_K \sum_p \langle n_{K,p} \rangle \hbar\omega_{K,p} \ , \tag{4.1}$$

where: represents the phonon thermal equilibrium occupancy with wavevector K and polarization p. The expression for is determined by the Planck distribution function (Lin & Zhigilei, 2007).

$$\langle n \rangle = \frac{1}{\exp(\hbar\omega/\tau) - 1} \ , \tag{4.2}$$

The symbol represents the average value in thermal equilibrium. Figure 4.1 displays a graph representing the variable (n) (Naumov & Musikhin, 2015).

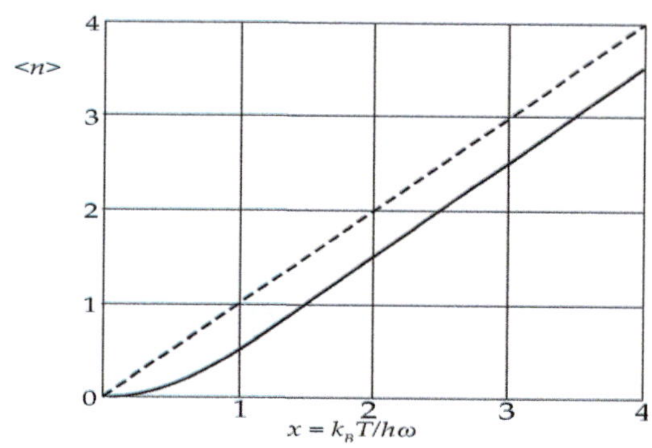

Figure 4.1. Planck distribution function plot. The occupancy of a state is roughly linear with respect to temperature at high temperatures.

Source: Abragam, A., Creative Commons License.

4.1.1. Planck Distribution

Let's examine a collection of identical harmonic oscillators that are in a state of thermal equilibrium. Calculating the ratio of oscillators in different quantum states of excitation can be done using the Boltzmann factor (Jagielski & Kutner, 2013).

Remember

Phonons are discrete units of lattice vibrations in solids, similar to photons for electromagnetic waves.

$$N_{n+1}/N_n = \exp(-\hbar\omega/\tau) \ , \qquad \tau \equiv k_B T \ , \tag{4.3}$$

The following method can be used to calculate the fraction of oscillators in the nth quantum state:

$$\frac{N_n}{\sum_{s=0}^{\infty} N_s} = \frac{\exp(-n\hbar\omega/\tau)}{\sum_{s=0}^{\infty} \exp(-s\hbar\omega/\tau)} \ . \tag{4.4}$$

We observe that an oscillator's average excitation quantum number is:

$$\langle n \rangle = \frac{\sum_s s \exp(-s\hbar\omega/\tau)}{\sum_s \exp(-s\hbar\omega/\tau)} \ . \tag{4.5}$$

The summations in (4.5) are:

$$\sum_s x^s = \frac{1}{1-x} \ ; \qquad \sum_s s x^s = x \frac{d}{dx} \sum_s x^s = \frac{x}{(1-x)^2} \ , \tag{4.6}$$

using the formula . Therefore, we can rephrase (4.5) as the distribution of Planck (Martin, 1991).

$$\langle n \rangle = \frac{x}{1-x} = \frac{1}{\exp(\hbar\omega/\tau) - 1} \ . \tag{4.7}$$

4.1.2. Normal Mode Enumeration

(4.1) and (4.2) yield the energy of a group of oscillators in thermal equilibrium with frequency :

$$U = \sum_K \sum_p \frac{\hbar\omega_{K,p}}{\exp(\hbar\omega_{K,p}/\tau) - 1} \ . \tag{4.8}$$

An integral can typically be used in place of the summation over K. Assume that the crystal exhibits modes in the frequency range of , for a given polarization p (Ivers et al., 2015).

Then the energy is:

$$U = \sum_p \int d\omega\, D_p(\omega) \frac{\hbar\omega}{\exp(\hbar\omega/\tau) - 1} . \tag{4.9}$$

The process of differentiation with regard to temperature yields the lattice heat capacity. Assuming : then yields:

$$C_{\rm lat} = k_B \sum_p \int d\omega\, D_p(\omega) \frac{x^2 \exp x}{(\exp x - 1)^2} . \tag{4.10}$$

Finding D(ω), or the number of modes per unit frequency range, is the main challenge. This function is known as the density of modes, or more widely, the density of states (Pannocchia et al., 2007).

4.1.3. States Density in One Dimension

Let's explore the boundary value problem related to the one-dimensional line vibrations (Figure 4.2) that consists of N + 1 particles separated by a distance a. We suppose:

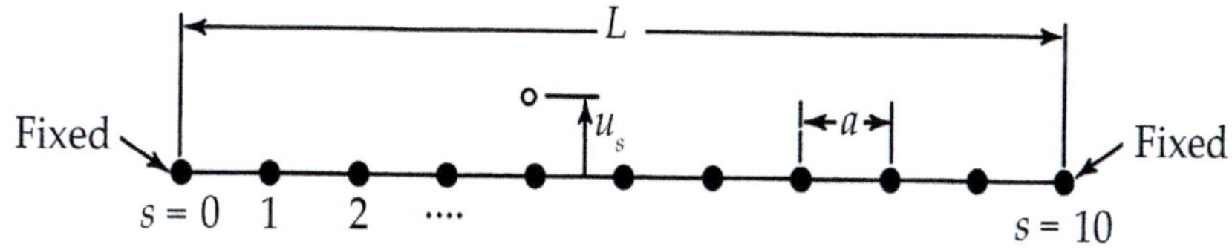

Figure 4.2. Illustration of elastic line of N+1 atoms, for N= 10, with the fixed end atoms, s=0 and s=10 as boundary conditions. For both transverse and longitudinal displacements, the displacements of particle in normal modes have the form sin sKa. At the end s=0, this form is naturally zero at the atom. To make the displacement zero at the end of s=10, we select K.

Source: Chikazumi, S., Creative Commons License.

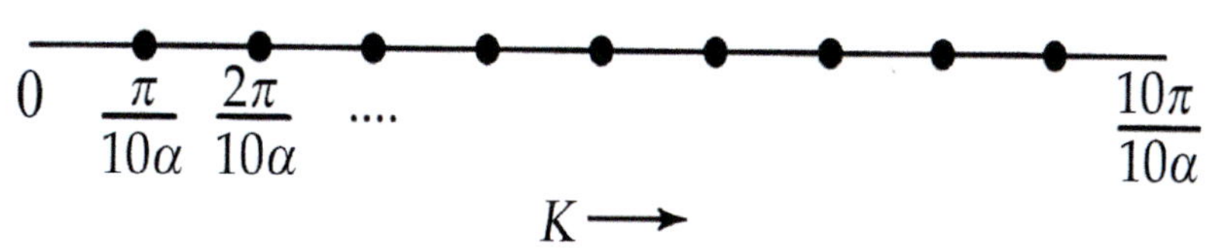

Figure 4.3. Illustration of the boundary condition for sin sKa = 0.

Source: Fulde, P., Creative Commons License.

Figure 4.3 the boundary condition sin sKa = 0 for s = 10 can be satisfied by choosing K = π/10a, 2π/10a, or 9π/10a, where 10a is the length of the line L. As of now, the figure is in K space. Instead of atoms, the dots denote the allowed values of K. The only particles allowed to move out of the N + 1 on the line are N — 1, and the N — 1 permits for K can be used to characterize their most common motion (Alexander & Orbach, 1982).

This quantization of K is independent of quantum physics and arises conventionally from the boundary constraints requiring the end atoms to be fixed, that the particles are kept motionless at the line's endpoints, s = 0 and s = N. Each common vibrational mode of polarization (p) is a standing wave, where us is the displacement of the particles:

$$u_s = u(0)\exp(-i\omega_{K,p}t)\sin sKa \quad , \qquad (4.11)$$

Similar to Figure 4.3, the fixed-end boundary conditions limit the wave vector K to the values:

$$K = \frac{\pi}{L},\ \frac{2\pi}{L},\ \frac{3\pi}{L},\ldots,\ \frac{(N-1)\pi}{L} \quad . \qquad (4.12)$$

The solution for K = π/L has:

$$u_s \propto \sin\,(s\pi a/L) \qquad (4.13)$$

and vanishes for s = 0 and s = N as required.

Since sin sw disappears for every atom, the solution for K = Nπ/L = π/a = has d sin sπ, which prevents any atom from moving. Consequently, in equation (12), N — 1 independent values of K are permitted. This number is the same as the number of particles that are allowed to move. A standing wave corresponds to each allowed value of K. For any interval ΔK = π/L in the one-dimensional line, there is a single mode. This suggests that there are no modes for K values larger than π/a. Thus, L/π K < π/a is the number of modes per unit range of K (Himpsel et al., 2001).

In the context of solid-state physics, it is worth noting that for every value of K, there exist three polarizations p. It is interesting to observe that out of these three, two are transverse while one is longitudinal, but only in one dimension. For wave vectors in specific crystal directions, the polarizations become simplified in three dimensions (Luther & Peschel, 1974).

Another method for counting modes is equally acceptable. In our analysis, we treat the medium as infinite, but we impose the condition that the solutions exhibit periodicity over a significant distance L, meaning that u(sa) = u(sa + L).

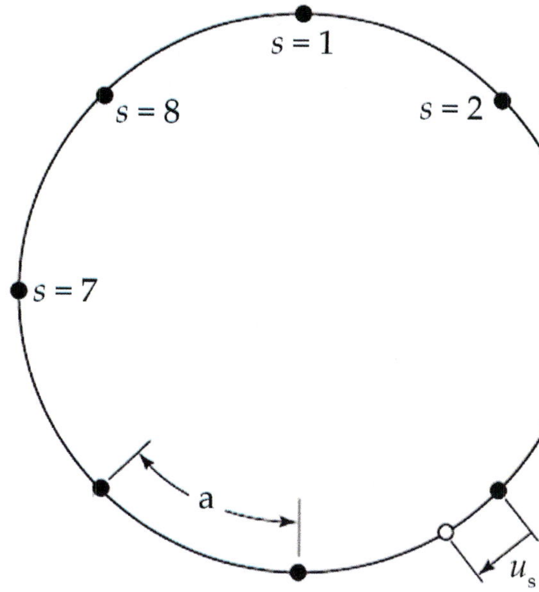

Figure 4.4. Illustration of constrained circular ring.

Source: Heeger, A. J, Creative Commons License.

Figure 4.4. Consider of N particles that are forced to slide around a ring. If there are elastic springs connecting the particles, they can oscillate. The displacement of an atom in a normal mode will take the form of either sin sKa or cos sKa; these are separate modes. NKa needs to be an integral multiple of 2w since the boundary requirement for the ring's geometrical periodicity is that = for all s. The independent values of K that are permitted for N = 8 are 0, 2π/8a, 4π/8a, 6π/8a, and 8π/8a. Because sin s0a = 0, the number K = 0 has no significance in the sine form (Giamarchi, 2004).

Because sin (s8πa/8a) = sin s π=0, the value 8w/8a only makes sense in the cosine form. For the eight particles, there are a total of eight modes since the sine and cosine modes accommodate the final three values of K. Similar to Figure 4.3's fixed-end boundary condition, the periodic boundary condition permits one admissible mode for each particle. If the periodic boundary condition is applied, there will be eight modes with values of K = 0, ±2 π/Na, ±4 π/Na, ±6 π/Na, and 8 π/Na, as stated in Equation (4.14) (Wagner et al., 2012), if the modes were represented in the complex form exp (isKa).

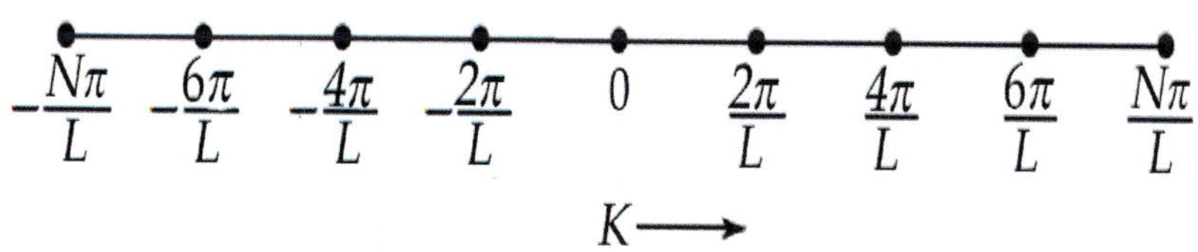

Figure 4.5. Illustration of 5 wave vector K allowed values for periodic boundary conditions on a linear lattice.

Source: Hewson, A. C, Creative Commons License.

Exploring the permissible wave vector values for a linear lattice with periodic boundary conditions is crucial when studying the behavior of a solid-state system. In this case, we are considering a lattice with a periodicity of N = 8 atoms on a line of length L. The K = 0 solution represents the uniform mode in this context. There are only eight allowed modes in this case. The displacements of the atoms are proportional to different exponential values, including 1, $\exp(\pm i\pi s/4)$, $\exp(\pm i\pi s/2)$, $\exp(\pm i3\pi s/4)$, and $\exp(i\pi s)$ (Figure 4.5).

For a large system, the periodic boundary conditions method (Figs. 4.4 and 4.5) does not fundamentally alter the physics of the problem. is the running wave solution. The K values that are permitted are:

$$K = 0 \ , \quad \pm\frac{2\pi}{L} \ , \quad \pm\frac{4\pi}{L} \ , \quad \pm\frac{6\pi}{L} \ , \ \ldots , \ \frac{N\pi}{L} \ . \tag{4.14}$$

This enumeration approach yields an equal number of modes (one for each mobile atom) as indicated by equation (12). However, we now have both positive and negative values of K, with a spacing of $\Delta K = 2\pi/L$ between consecutive K values. Periodic boundary conditions result in $L/2\pi$ modes per unit range of K within the interval $-\pi/a < K \leq \pi/a$, and no modes outside of this interval. Figure 4.6 illustrates the configuration of a two-dimensional lattice (Lang et al., 2012).

$D(\omega)$ must be known in order to calculate the number of modes per frequency range for a given polarization. The number of modes $D(\omega)\, d\omega$ at ω in one dimension is given by

$$D_1(\omega)\, d\omega = \frac{L}{\pi}\frac{dK}{d\omega} d\omega = \frac{L}{\pi} \cdot \frac{d\omega}{d\omega/dK} \ . \tag{4.15}$$

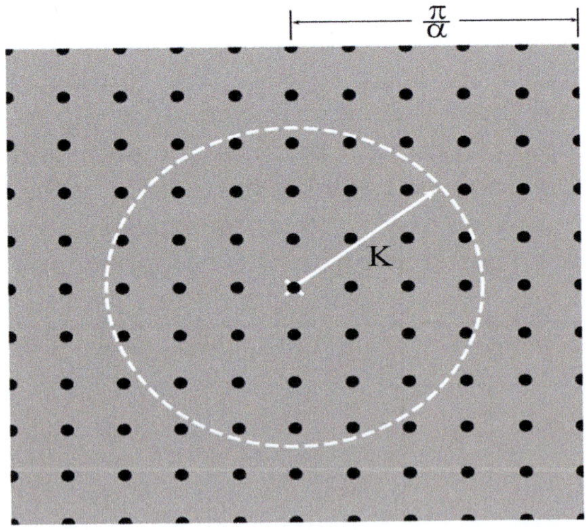

Figure 4.6. Illustration of allowed values of the phonon wave vector K in Fourier space for a square lattice.

Source: Kondo, J., Creative Commons License.

The group velocity, , can be derived from the dispersion relation that relates ω and K. A singularity occurs in (ω) whenever the dispersion relation is horizontal, indicating that the group velocity is zero.

4.1.4. States Density in Three Dimensions

primitive cells in a cube of side L are subjected to periodic boundary conditions, which yield K depending on a certain condition.

$$\exp[i(K_x x + K_y y + K_z z)] \equiv \exp\{i[K_x(x+L) + K_y(y+L) + K_z(z+L)]\} \;, \tag{4.16}$$

Hence:

$$K_x, K_y, K_z = 0 \;; \quad \pm\frac{2\pi}{L} \;; \quad \pm\frac{4\pi}{L} \;; \quad \cdots; \quad \frac{N\pi}{L} \;. \tag{4.17}$$

Thus, in K space, there is only one permissible value of K per volume .

$$\left(\frac{L}{2\pi}\right)^3 = \frac{V}{8\pi^3} \tag{4.18}$$

For each polarization and branch, K values are computed per unit volume of K space. The formula V = can be used to determine the specimen volume (Kobayashi et al., 2014).

Equation (4.18) can be used to find the total number of modes having a wavevector less than K. It is equal to the volume of a sphere with radius K multiplied by .

Thus:

$$N = (L/2\pi)^3(4\pi K^3/3) \tag{4.19}$$

for every polarization type. With any polarization, the density of states equals:

$$D(\omega) = dN/d\omega = (VK^2/2\pi^2)(dK/d\omega) \;. \tag{4.20}$$

4.1.5. Debye Model for Density of States

Within the Debye approximation, the velocity of sound is considered to be constant for every type of polarization, resembling that of a classical elastic continuum. The mathematical expression of the dispersion relation is provided (Schliesser & Woodfield, 2015).

$$\omega = vK \;, \tag{4.21}$$

The density of states (4.20) becomes:

$$D(\omega) = V\omega^2/2\pi^2 v^3 \ . \tag{4.22}$$

The number of acoustic phonon modes in the specimen is equal to the number of primitive cells, N. The cutoff frequency is calculated using equation (4.19) as:

$$\omega_D^3 = 6\pi^2 v^3 N/V \ . \tag{4.23}$$

A cutoff wave vector in K space corresponds to this frequency:

$$K_D = \omega_D/v = (6\pi^2 N/V)^{1/3} \ . \tag{4.24}$$

In the Debye model, wave vectors bigger than are not permitted. The number of modes with a value of K less than or equal to is equal to the number of degrees of freedom in a monatomic lattice (Duke & Mahan, 1965).

The thermal energy (9) is determined by:

$$U = \int d\omega\, D(\omega)\langle n(\omega)\rangle\hbar\omega = \int_0^{\omega_D} d\omega \left(\frac{V\omega^2}{2\pi^2 v^3}\right)\left(\frac{\hbar\omega}{e^{\hbar\omega/\tau} - 1}\right) , \tag{4.25}$$

for any kind of polarization. Given the assumption of the phonon velocity being independent of polarization, we can multiply by three to obtain the desired result.

$$U = \frac{3V\hbar}{2\pi^2 v^3}\int_0^{\omega_D} d\omega \frac{\omega^3}{e^{\hbar\omega/\tau} - 1} = \frac{3Vk_B^4 T^4}{2\pi^2 v^3 \hbar^3}\int_0^{x_D} dx \frac{x^3}{e^x - 1} , \tag{4.26}$$

where: and:

$$x_D \equiv \hbar\omega_D / k_B T \equiv \theta/T \ . \tag{4.27}$$

The Debye temperature θ is defined here in terms of , which is determined by (4.23). θ can be expressed as:

$$\theta = \frac{\hbar v}{k_B} \cdot \left(\frac{6\pi^2 N}{V}\right)^{1/3} , \tag{4.28}$$

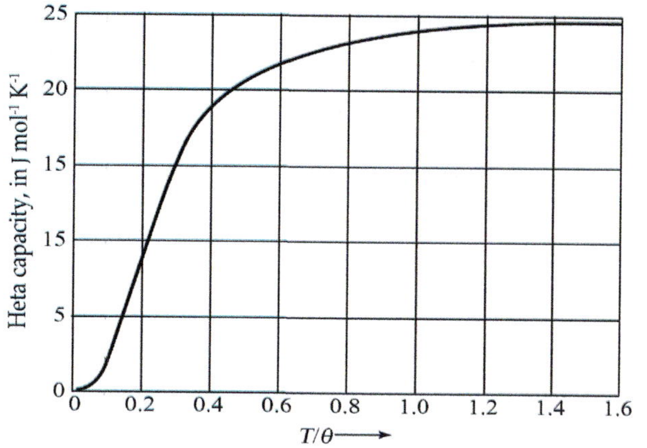

Figure 4.7. Illustration of a solid's heat capacity (CV) using the Debye approximation.

Source: Kubo, R, Creative Commons License.

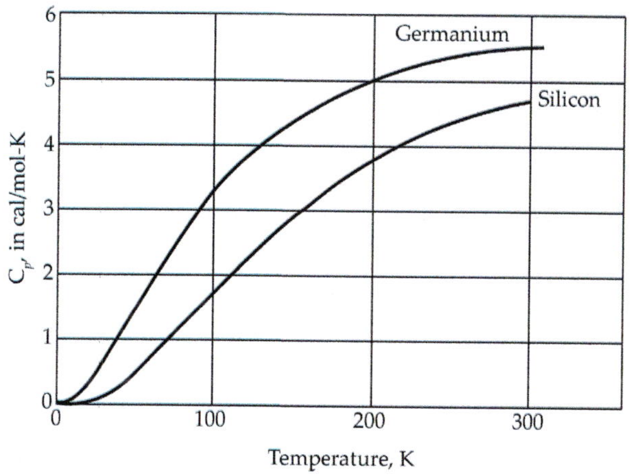

Figure 4.8. Representation of germanium and silicon's heat capacities. Take a note of the drop in colder temps. Multiply by 4.186, to convert a value in cal/mol-K to J/mol-K.

Source: Mattis, D. C., Creative Commons License.

so that the total phonon energy is:

$$U = 9Nk_BT\left(\frac{T}{\theta}\right)^3 \int_0^{x_D} dx \frac{x^3}{e^x - 1} , \tag{4.29}$$

where: and N is the number of atoms in the specimen.

The simplest method for determining the heat capacity is to differentiate the middle expression of (26) with regard to temperature (Thoss et al., 2001). Then:

$$C_V = \frac{3V\hbar^2}{2\pi^2 v^3 k_B T^2} \int_0^{w_D} dw \frac{\omega^4 e^{\hbar w/\tau}}{(e^{\hbar w/\tau} - 1)^2} = 9Nk_B\left(\frac{T}{\theta}\right)^3 \int_0^{x_D} dx \frac{x^4 e^x}{(e^x - 1)^2} . \tag{4.30}$$

In Figure 4.7, the Debye heat capacity is plotted. The heat capacity gets closer to the traditional figure of 3NkB when T ≫ θ. Plots of the measured values for germanium and silicon are shown in Figure 4.8.

4.1.6. Debye Law

By allowing the upper limit to reach infinity, we can approximate (4.29) at very low temperatures. We have:

$$\int_0^\infty dx \frac{x^3}{e^x - 1} = \int_0^\infty dx\, x^3 \sum_{s=1}^{\infty} \exp(-sx) = 6 \sum_1^\infty \frac{1}{s^4} = \frac{\pi^4}{15}, \tag{4.31}$$

where typical tables find the sum over :

$$C_V \cong \frac{12\pi^4}{5} Nk_B \left(\frac{T}{\theta}\right)^3 \cong 234\, Nk_B \left(\frac{T}{\theta}\right)^3, \tag{4.32}$$

This is the approximation for Debye . Plotting of argon experimental findings is shown in Figure 4.9.

At sufficiently low temperature the approximation is quite good; that is, when only long wavelength acoustic modes are thermally excited. These are just the modes that may be treated as an elastic continuum with macroscopic elastic constants. The energy of the short wavelength modes (for which this approximation fails) is too high for them to be populated significantly at low temperatures. We understand the result by a simple argument (Fig. 4.10).

At a low temperature T, only the lattice modes with will be significantly excited. The activation of these modes will be roughly classical, each with an energy near as shown in Figure 4.1. The fraction of the allowed volume in K space occupied by the excited modes can be approximated by either or . Here, represents a "thermal" wavevector defined as , and K_D is the Debye cutoff wavevector (Karpov & Parshin, 1985).

Therefore, the fraction occupied in the total volume of K space is given by (. There are approximately excited modes, with each mode having an energy of . The energy is approximately (~3N and the heat capacity (~12N.

The approximation is applicable to real crystals only at very low temperatures. To achieve a sufficiently pure behavior, it may be essential to maintain a temperature below T = θ/50.

4.1.7. Density of States Model by Einstein

Let's examine a system consisting of N oscillators with identical frequencies ω0, all confined to one dimension. The Einstein density of states is given by the function , where the delta function is centered at . The system possesses thermal energy (Dimitrov et al., 2010).

$$U = N\langle n\rangle\hbar\omega = \frac{N\hbar\omega}{e^{\hbar\omega/\tau} - 1}\ , \tag{4.33}$$

with now written in place of , for convenience:

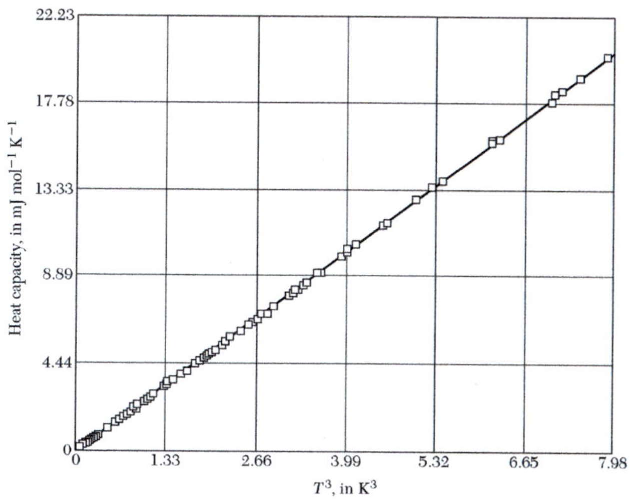

Figure 4.9. Illustration of plotting solid argon's low temperature heat capacity vs T^3. The experimental results show good agreement with the Debye T^3 rule with K in this temperature range.

Source: Morrish, A. H., Creative Commons License.

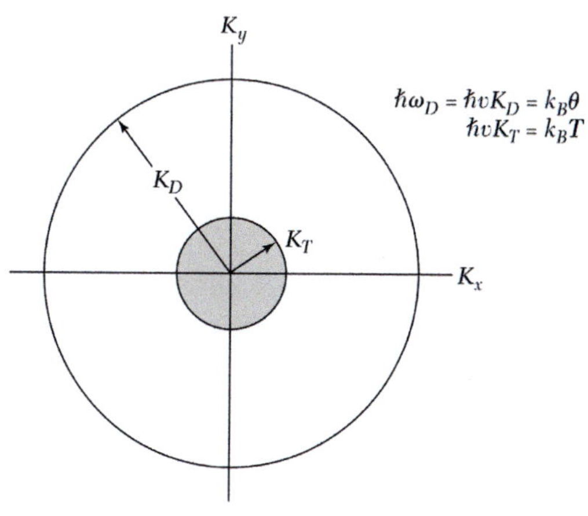

Figure 4.10. Assuming that all phonon modes with wave vectors lower than KT have the classical thermal energy kBT and that no modes are activated between KT and Debye cutoff KD allows us to obtain a qualitative explanation of the Debye T3 law.

Source: Morrish, A. H., Creative Commons License.

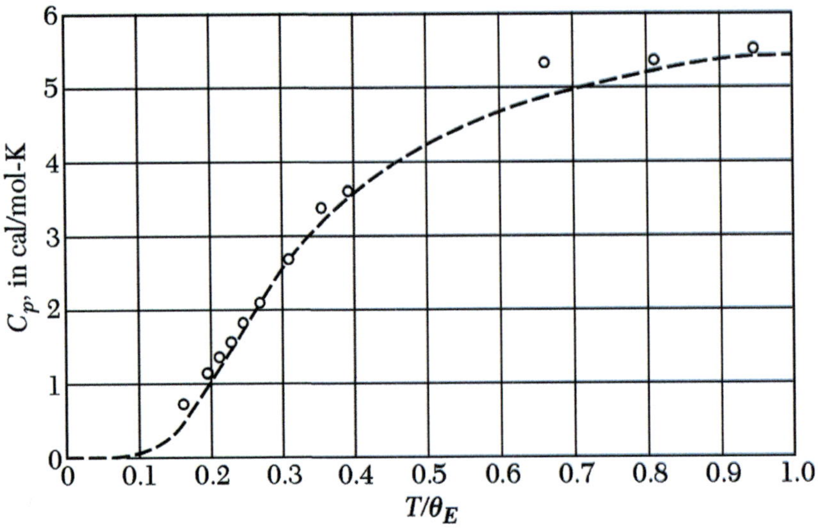

Figure 4.11. Illustration of diamond heat capacity experimental measurements are compared to those computed using the oldest quantum (Einstein) model, using θE = ⊠ω/ = 1320 K as the characteristic's temperature. Multiply by 4.186 to convert to J/moldeg.

Source: Nolting, W., Creative Commons License.

Figure 4.11 shows the oscillators' heat capacity plotted.

$$C_V = \left(\frac{\partial U}{\partial T}\right)_V = Nk_B \left(\frac{\hbar\omega}{\tau}\right)^2 \frac{e^{\hbar\omega/\tau}}{(e^{\hbar\omega/\tau} - 1)^2}, \tag{4.34}$$

This is the equation that Einstein used in 1907 to calculate the effect of N identical oscillators on a solid's heat capacity. In a three-dimensional setting, the variable N is multiplied by 3N, as each oscillator now has three modes. The upper bound for the temperature at which reaches its maximum value is given by , which is commonly referred to as the Petit and Dulong value (Hermann et al., 2003).

At low temperatures, the quantity (4.34) drops exponentially with the term . On the other hand, experimental representation of the contribution of phonon follows a dependence, as explained by the Debye model discussed above. The Einstein model is commonly employed to estimate the optical phonon component of the phonon spectrum (Xiao et al., 2016).

4.1.8. General Result for D(w)

Using the phonon dispersion relation ω (K), we aim to derive a universal equation for D (ω), representing the quantity of states inside a specified frequency interval. The total number of permissible K values that satisfy the condition that the phonon frequency lies within the range of ω to ω + dω is:

$$D(\omega)\,d\omega = \left(\frac{L}{2\pi}\right)^3 \int_{\text{shell}} d^3K \ ,$$ (4.35)

Two surfaces in K space determine the volume of the shell, across which the integral is taken when the phonon frequency remains constant (Buchert, 2000). One surface has a frequency of ω, while the second surface has a frequency of ω + dω.

The main challenge lies in determining the volume of this shell. Let us examine a specific area (Figure 4.12) on the surface in K space that corresponds to the chosen constant. We will refer to this area as , with a frequency denoted as ω.

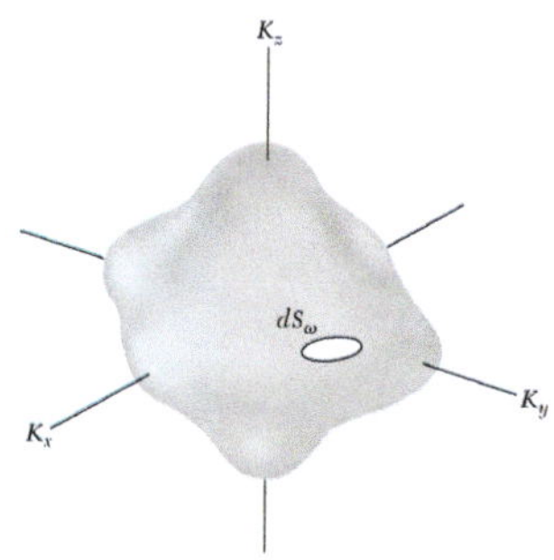

Figure 4.12. Illustration area on a surface in K space with constant frequency.

Source: Rado, G. T, Creative Commons License.

The volume element bounded by the constant frequency surfaces ω and can be described as a right cylinder with a base area of and a height of :

$$\int_{\text{shell}} d^3K = \int dS_\omega dK_\perp \ ,$$ (4.36)

Here, , represents the perpendicular distance between the surfaces ω constant and ωconstant (as shown in Figure 4.13). The magnitude of the perpendicular component of the gradient, denoted as , will exhibit variation across different points on the surface (Kijima, 1989).

The gradient of ω, which is ∇Kω, is perpendicular to the surface ω constant, and represents the frequency difference between the two connected surfaces by :

$$|\nabla_{\mathbf{K}}\omega|\, dK_\perp = d\omega \ ,$$

Therefore, the volume element can be expressed as:

$$dS_\omega\, dK_\perp = dS_\omega \frac{d\omega}{|\nabla_{\mathbf{K}}\omega|} = dS_\omega \frac{d\omega}{v_g} \ ,$$

A phonon's group velocity magnitude is determined by the expression . Regarding equation (4.35), we have:

$$D(\omega)\, d\omega = \left(\frac{L}{2\pi}\right)^3 \int \frac{dS_\omega}{v_g}\, d\omega \ .$$

After dividing both sides by and denoting the volume of the crystal as V = , we obtain the density of states (Conturo et al., 1996).

$$D(\omega) = \frac{V}{(2\pi)^3} \int \frac{dS_\omega}{v_g} \ . \qquad (4.37)$$

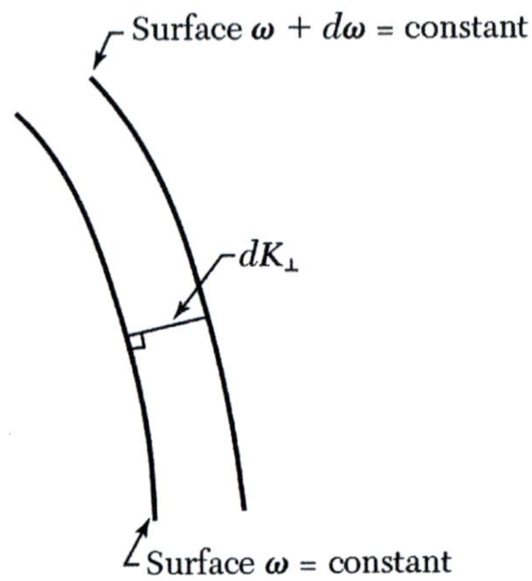

Figure 4.13. Illustration of the quantity represent the distance between two surfaces in K space, with one surface at frequency ω and the other at frequency ω + dω.

Source: Rigamonti, A., Creative Commons License.

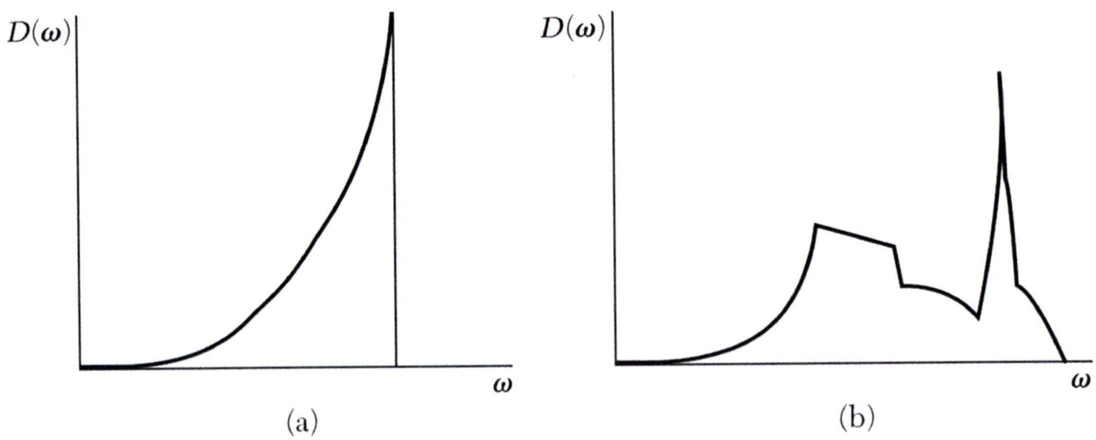

Figure 4.14. Density states for two different systems; a real crystal structure and (b) the debye solid as a function of frequency. For small ω, the crystal's spectrum begins as but discontinues appear at unique points.

Source: Rigamonti, A., Creative Commons License.

The integral is calculated over the surface area ω, which remains constant in K space. This finding relates to a specific aspect of the dispersion relation. This result can also be applied in electron band theory. The contribution to D (ω) from locations where the group velocity is zero is given special attention. The distribution function experiences singularities as a result of these pivotal points, most especially Van Hove singularities. (Figure 4.14) (Cooper & Simmons, 1977).

4.2. ANHARMONIC CRYSTAL INTERACTIONS

Lattice vibrations have been discussed in terms of quadratic interatomic displacements because the potential energy is constrained. This theory examines harmonic principles and their effects:

- One important characteristic of lattice waves is their stability over time. Unlike other types of waves, a single lattice wave does not degrade or change form as time passes. Additionally, it is worth noting that two lattice waves do not interact with each other.
- Thermal expansion does not occur.
- The elastic constants that are adiabatic and isothermal are same.
- Temperature and pressure have no effect on the elastic constants.
- When temperatures rise over θ, the heat capacity remains constant (Cowley & Cowley, 1965).

None of these consequences are accurately satisfied in real crystals. One possible explanation for the deviations is the failure to consider anharmonic terms in the interatomic displacements that go beyond quadratic. We explore certain fundamental aspects of anharmonic effects (Geller et al., 1972).

Anharmonic effects are fascinatingly revealed by experiments demonstrating the intriguing interactions of phonons, where two phonons combine to form a third phonon at a frequency . The lattice potential energy's third-order terms give birth to three-phonon processes (Young et al., 2015).

The concept of phonon interaction in physics can be explained in a simple way: the existence of a single phonon leads to a repetitive elastic strain that, due to the anharmonic interaction, alters the elastic constant of the crystal in both space and time. Another phonon detects the variation in the elastic constant and is subsequently deflected, resulting in the generation of a third phonon, similar to the effect of a mobile three-dimensional grating (Dragunov et al., 1997).

Remember

Anharmonic interatomic interactions within a crystal lattice play a role in phenomena such as thermal expansion and have an impact on the specific heat capacity at elevated temperatures.

4.3. THERMAL EXPANSION

One way to understand thermal expansion is to look at how anharmonic components in the potential energy affect the average distance between two atoms in a classical oscillator at a certain temperature, T. When c, g, and f are all positive, the atoms' displacement (x) from their equilibrium separation can be used to indicate the potential energy of the atoms at zero temperature (Blackman, 1957).

$$U(x) = cx^2 - gx^3 - fx^4, \tag{4.38}$$

The asymmetry of atoms' mutual repulsion is captured by the x^3 component, and the softening of vibrations at larger amplitudes is described by the x^4 term. Although the minimum at x = 0 is not an absolute minimum, the form provides a good enough representation of an interatomic potential for tiny oscillations (Chen et al., 2015).

In order to determine the average displacement, we utilize the Boltzmann distribution function. This function assigns weights to different values of x based on their thermodynamic probability.

$$\langle x \rangle = \frac{\int_{-\infty}^{\infty} dx\, x \exp[-\beta U(x)]}{\int_{-\infty}^{\infty} dx \exp[-\beta U(x)]}$$

with . For displacements such that the anharmonic terms in the energy are small in comparison with kBT, we may expand the integrands as (Schapery, 1968).

$$\int dx\, x \exp(-\beta U) \cong \int dx[\exp(-\beta cx^2)](x + \beta gx^4 + \beta fx^5) = (3\pi^{1/2}/4)(g/c^{5/2})\beta^{-3/2};$$
$$\int dx \exp(-\beta U) \cong \int dx \exp(-\beta cx^2) = (\pi/\beta c)^{1/2}, \tag{4.39}$$

Hence the thermal expansion is:

$$\langle x \rangle = \frac{3g}{4c^2} k_B T \tag{4.40}$$

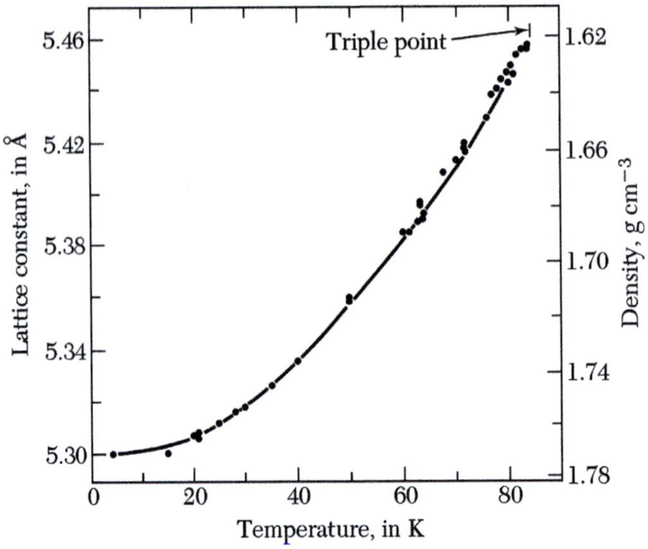

Figure 4.15. Illustration of lattice constant of solid argon varies as a function of temperature, measured in Kelvin (K).

Source: Van Vleck, Creative Commons License.

Solid argon lattice constant observations are shown in Figure 4.15. Curve gradient is proportional to thermal expansion coefficient. Problem 5 predicts that the expansion coefficient approaches 0 as temperature approaches zero (Evans et al., 1996). At the lowest level, thermal expansion is solely affected by the anti-symmetric term ⟦in U(x), and not by the symmetric term .

4.4. THERMAL CONDUCTIVITY

In a steady state, the rate at which heat moves through a long rod with a temperature gradient (dT/dx) gives the thermal conductivity coefficient K of that material (Padture & Klemens, 1997).

$$j_U = -K\frac{dT}{dx},$$

(4.41)

where is the thermal energy flux, or the amount of energy transferred in a unit of time over a unit of area.

This structure suggests that thermal energy transmission occurs randomly. The energy diffuses throughout the specimen, colliding with it frequently, rather than entering one end and traveling straight forward (ballistically) to the other (Smith et al., 2013).

The thermal flux expression would only depend on the temperature differential ΔT between the ends of the specimen, independent of its length, rather than the temperature gradient, if the energy were to flow through the specimen without any deviation. Because the conductivity process is random, the expression for the thermal flux includes the temperature gradient as well as, as we shall see, a mean free path (Slack et al., 1987).

The expression for thermal conductivity that follows is derived from the kinetic theory of gases:

$$K = \frac{1}{3}Cv\ell$$

(4.42)

where is the mean free path of a particle between collisions; is the average particle velocity; and is the heat capacity per unit volume. Debye was the first to apply this conclusion to explain dielectric material thermal conductivity, where is the phonons' heat capacity, their velocity, and their mean free path (Zhang et al., 2002).

We present the fundamental kinetic theory that results in equation (4.42). The particle flux in the x direction is given by the equation 1n(| v |), where n represents the concentration of molecules. There is an equal and opposing flow of particles in the opposite direction when everything is in equilibrium. The average value is shown by the ⟨...⟩ symbols (Rosenberg, 1955).

When a particle transitions from a place with a temperature of T + ΔT to an area with a temperature of T, it will emit an energy quantity equivalent to c ΔT, where c denotes the particle's heat capacity. The change in temperature (ΔT) across the length of a particle's free path can be expressed as:

$$\Delta T = \frac{dT}{dx}\ell_x = \frac{dT}{dx}v_x\tau \ ,$$

where: τ represents the meantime interval between collisions.

Consequently, the overall flow of energy (from both directions of the flow of particles) is thus:

$$j_U = -n\langle v_x^2\rangle c\tau\frac{dT}{dx} = -\tfrac{1}{3}n\langle v^2\rangle c\tau\frac{dT}{dx} \ . \tag{4.43}$$

If the velocity, denoted by v, remains constant for phonons, we can express equation (4.43) as (Francl & Kingery, 1954).

$$j_U = -\tfrac{1}{3}Cv\ell\frac{dT}{dx} \ , \tag{4.44}$$

where: C nc; and . Thus .

4.4.1. Phonon Gas Thermal Resistivity

Two basic elements govern the phonon mean free path: geometric scattering and scattering from other phonons. Collisions between distinct phonons would not occur in a situation in which the interactions between atoms were just harmonic. Thus, only phonon collisions with the crystal border and lattice defects would define the mean free path. In certain cases, these impacts are more noticeable.

The mean free path is constrained by the interaction between different phonons in anharmonic lattices. The precise states of the anharmonic system are distinct from those of pure phonons (Lortz et al., 2006).

At elevated temperatures, the theory of anharmonic coupling predicts that 9 is directly proportional to 1/T in terms of thermal resistivity. This aligns with multiple experimental observations. The correlation between this reliance becomes evident when we take into account the extent to which a specific phonon can engage with other phonons: at elevated temperatures, the overall quantity of activated phonons is directly proportional to the temperature, T. The frequency at which phonons collide is directly proportional to the total quantity of phonons available for collision ($l \propto 1/T$). In order to establish a thermal conductivity, it is necessary for the crystal to have mechanisms that can locally bring the distribution of phonons into thermal equilibrium. In the absence of these mechanisms, it is impossible to assume that the phonons at one end of the crystal are in thermal equilibrium at temperature T2, while the phonons at the other end are in equilibrium at temperature T1 (Hatke et al., 2011). In order to ensure an effective approach, it is crucial to not only have a method for restricting the average distance traveled by particles, but also to establish a localized distribution of phonons in thermal equilibrium. Collisions between phonons and static imperfections or crystal boundaries do not lead to thermal equilibrium on their own. This is due to the fact that individual phonon energy is not changed by these collisions. Stated otherwise, the incident phonon frequency and the dispersed phonon frequency are constant (Allen et al., 1994).

Did you know?

The high heat conductivity of diamond is due to its robust carbon-carbon bonds, which facilitate effective phonon conduction inside the lattice.

It is interesting to note that, for a specific reason, a three-phonon collision process does not result in equilibrium: the phonon gas's overall momentum does not change.

$$\mathbf{K}_1 + \mathbf{K}_2 = \mathbf{K}_3 \tag{4.45}$$

At temperature T, a stable distribution of phonons can propagate through the crystal with a consistent drift velocity, unaffected by three-phonon collisions of the form (4.45).

$$\mathbf{J} = \sum_{\mathbf{K}} n_{\mathbf{K}} \hbar \mathbf{K} \tag{4.46}$$

In these collisions, the momentum of the phonon is preserved, as the change in J, represented by , equals zero upon contact. nK represents the quantity of phonons with a specific wave vector K (Milman et al., 1992). When dealing with a distribution where J is not equal to zero, it is important to note that collisions like (4.45) do not have the ability to achieve full thermal equilibrium as they do not alter J. The distribution will move along the rod without changing J when hot phonons are first propagating along a rod with a non-zero J. Heat resistance is therefore nonexistent. The problem, as shown in Figure 4.16, is similar to the case of gas molecules colliding in a straight tube with non-frictional walls (Jaffe et al., 2019).

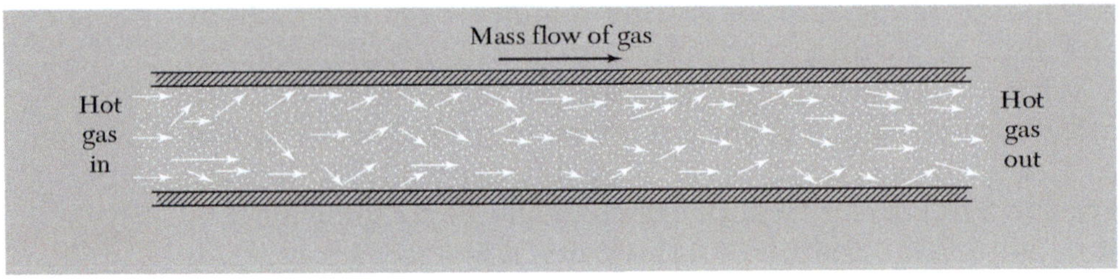

Figure 4.16. Illustration of mass flow of gas.

Source: White, R. M., Creative Commons License.

Gas molecules drifting down a long, open tube with frictionless walls in an equilibrium state. Elastic collision processes between gas molecules preserve the gas's momentum and energy flux since each collision results in a constant collision energy and center of mass velocity. As a result, energy moves from the left to the right without being driven by a change in temperature. Thus, thermal conductivity is infinite and thermal resistivity is zero (Lu et al., 2018) (Figure 4.17).

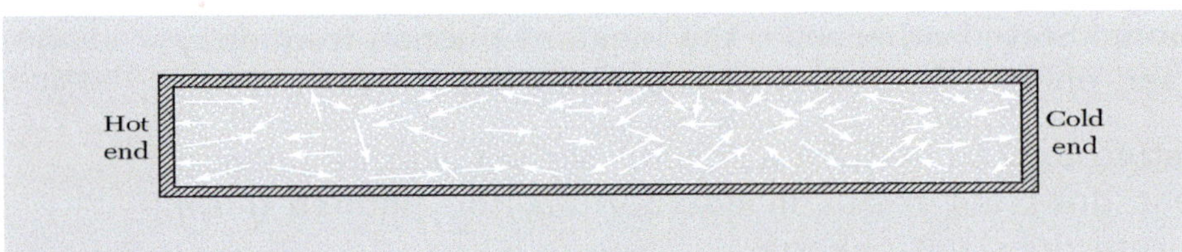

Figure 4.17. Illustration of a situation where no mass flow is permitted.

Source: Andreani, L. C., Creative Commons License.

In the context of gases, thermal conductivity is typically defined as the ability of a gas to conduct heat when there is no movement of mass. In this scenario, the tube is sealed at both ends, thereby blocking the passage of molecules in or out. When there is a difference in temperature between two regions, the pairs of particles that collide with higher-than-average speeds in their center of mass will generally go towards the right, while those with lower-than-average velocities would tend to move towards the left. A small gradient in concentration, with a higher concentration on the right side, will be established in order to achieve a balance in mass transport while facilitating the transfer of energy from the hot end to the cool end (Zudov et al., 2001).

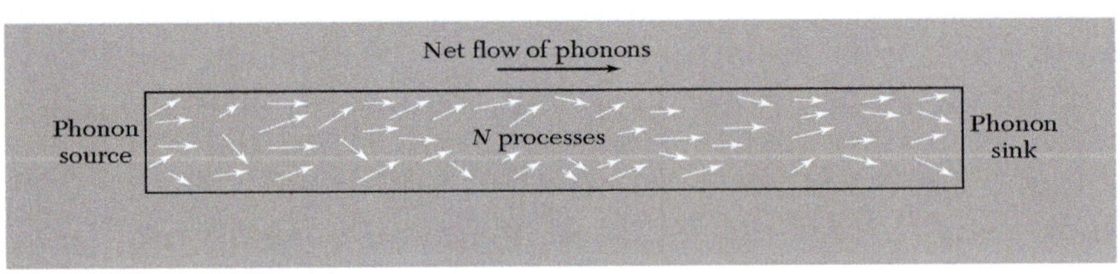

Figure 4.18. Illustration of net flow of phonons.

Source: Bilz, H., Creative Commons License.

By illuminating the left end of a crystal with a lamp, we can selectively generate phonons primarily at that end. Phonons will then start to flow in a directed manner in the direction of the crystal's right end. If there are only N processes occurring, where , the momentum of the phonon flux remains unchanged throughout collisions, and a certain amount of phonon flux will continue along the crystal's length. When phonons reach the right end, there is a potential to efficiently convert most of their energy into radiation, creating a mechanism for absorbing the phonons. In (a), the thermal resistance is negligible (Durczewski et al., 2020) (Figure 4.18).

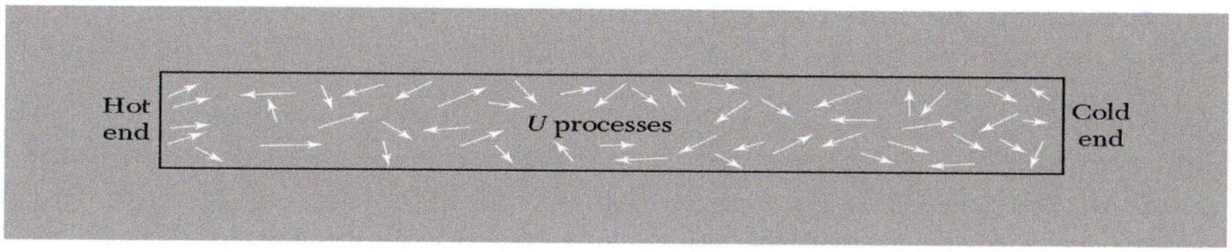

Figure 4.19. Illustration of mass flow form hot end to cold end.

Source: Brillouin, L., Creative Commons License

Each collision occurrence in U processes involves a significant alteration in phonon momentum. The phonon flux will decrease significantly as we progress towards the right. The ends have the ability to serve as both sources and sinks. Energy transport occurs when there is a difference in temperature, similar to scenario (b) (Williams et al., 1983) (Figure 4.19).

4.4.2. Umklapp Processes

Thermal resistivity is the result of significant three-phonon reactions that violate the conservation law , where K is conserved. Instead, they adhere to a different form (Moon et al., 2002).

$$\mathbf{K}_1 + \mathbf{K}_2 = \mathbf{K}_3 + \mathbf{G} \ , \quad (4.47)$$

G represents a reciprocal lattice vector. The processes, which were discovered by Peierls, are referred to as mklapp processes. It is important to note that G can be present in all momentum conservation equations in crystals. Energy is conserved in all permissible processes described by equations (4.46) and (4.47) (Sussmann & Thellung, 1963).

We have seen crystal wave interaction events where the wave vector change is a reciprocal lattice vector instead of zero. Such processes can arise in periodic lattices. For phonons, the evidence is particularly strong since the significant phonon K's are restricted to the first Brillouin zone. This implies that each additional K created during a collision needs to be moved by adding a G to the first zone (Maznev & Wright, 2014).

Anmklapp process (G ≠ 0) can produce a phonon with positive from the collision of two phonons, both of which have negative values of . U processes are another name for Umklapp processes. Normal processes, also referred to as N processes (Jauernik et al., 2018), are collisions in which G = 0.

Since ., all phonon modes are activated at high temperatures T > θ. This means that U processes will account for a significant fraction of all phonon collisions, with a correspondingly large shift in collision momentum. In this particular area of study, it is possible to determine the thermal resistivity without differentiating between U and N processes. Taking into account the earlier analysis of nonlinear effects, it is expected that a lattice thermal resistivity of ∞ T will be observed at elevated temperatures (Anderson & Lapeyre, 1976).

Umklapp requires about of energy to occur, where θ is the typical temperature and is the Boltzmann constant. This is due to the fact that for the collision (4.47) to occur, both phonons, represented by the letters , , must have wave vectors on the order of .

If both phonons have low wave vectors, suggesting minimal energy, their collision cannot generate a phonon outside the first zone. Energy conservation is a fundamental requirement for the mklapp process, just as it is for other processes (Westphal & Goldmann, 1983).

The Boltzmann factor signifies the quantity of suitable phonons possessing the high energy 1k θ required at low temperatures should change approximately as exp(—θ/2T). The exponential form and the experiment accord quite well. Ultimately, the phonon mean free path mentioned in equation (4.42) specifically refers to the mean free path associated with phonon collisions involving umklapps, rather than encompassing all phonon collisions (Maebashi & Fukuyama, 1998).

4.4.3. Imperfections

Geometrical effects can also impose limitations on the mean free path. It is crucial to consider various factors such as crystal boundary scattering, the isotopic masses distribution in chemical impurities, lattice flaws, natural chemical elements, and amorphous structures (Stapleton & Subrahmanyam, 1977). At low temperatures, the width limits the value of . At that point, the specimen's dimensions determine the thermal conductivity, and the mean free path become similar to test specimen width. Biermasz and de Haas made this discovery. The size effect is responsible for the sudden drop in pure crystal thermal conductivity at low temperatures (Xu & Bogers, 2024).

As seen in Figure 4.18, the size effect limits thermal conductivity at low temperatures, replacing mklapp process. It is reasonable to assume that the phonon mean free path will be constant and approximately equivalent to the specimen's diameter D (Schafer & Peköz, 1998).

$$K \approx CvD \ . \qquad (4.48)$$

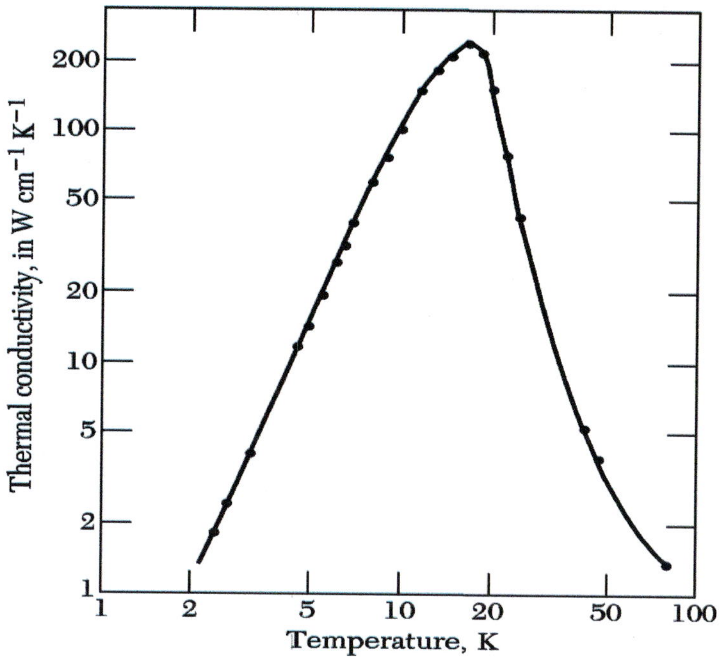

Figure 4.20. Illustration of highly purified crystal of sodium fluoride thermal conductivity.

Source: Brillouin, L., Creative Commons License.

Did you know?

Polycrystalline materials' mechanical and thermal properties depend on phonon interactions with imperfections such dislocations and grain boundaries

The heat capacity, or C, is the only temperature-dependent quantity on the right. At low temperatures, it varies as . At low temperatures, we anticipate that the thermal conductivity will change as . The size effect appears as soon as the diameter of the specimen and the phonon mean free path are equivalent (Wilson, 1942). (Figure 4.20).

The thermal conductivities of dielectric crystals can approach those of metals. One of the highest conductivity values is found in synthetic sapphire (), which is close to 200 W c at 30K. Sapphire's maximum thermal conductivity is higher than copper's maximum of 100 W c. On the other hand, at 1.8 K (Lukas et al., 2005), the conductivity of metallic gallium is 845 W c.

The distribution of chemical element isotopes in an otherwise flawless crystal frequently serves as a key mechanism for phonon scattering. An elastic wave observes that the periodicity of the density is disrupted by the random distribution of isotopic mass. In certain materials, the significance of phonon scattering by isotopes is equal to that of scattering by other phonons (Chen et al., 1999).

ACTIVITY 4.1

Objective: To investigate the concept of phonons and their role in determining thermal properties of solids.

Materials Needed:

- Computer with internet access for research;
- Pen and paper for notes.

Steps:

- **Research:** Use online resources or textbooks to gather information on phonons. Focus on understanding what phonons are, how they relate to lattice vibrations in solids, and their significance in thermal conductivity, heat capacity, and thermal expansion.
- **Compare Models:** Compare the Debye and Einstein models for the density of states in solids. Note their assumptions, mathematical formulations, and predictions for different temperature regimes.
- **Case Studies:** Research specific examples where phonon behavior and thermal properties play a crucial role. Examples could include materials with high thermal conductivity (like diamond) or those with interesting thermal expansion characteristics (like alloys).
- **Discussion:** Discuss with classmates or colleagues the implications of phonon behavior in real-world applications. Consider how advancements in understanding phonons could impact fields such as materials science, engineering, and technology.
- **Reflection:** Write a short reflection on what you have learned about phonons and their importance in the thermal properties of solids. Include any insights gained from comparing theoretical models and exploring case studies.

SUMMARY

- The chapter explores the basic ideas underlying the behavior of phonons, or lattice vibrations, in solids. The first two topics covered are the Planck distribution and phonon heat capacity, which are crucial to comprehending the distribution of energy among vibrational modes. A theoretical basis for characterizing the vibrational spectrum of solids is provided by the listing of normal modes and the computation of the density of states in one and three dimensions.
- The chapter presents the Debye model, which provides an estimation of the density of states and predicts phenomena such as the Debye T3 rule. This law is important for comprehending heat capacity at low temperatures. In addition, the Einstein model provides an alternate method for expressing the density of states in solids.
- Anharmonic crystal interactions are also discussed, which become important at higher temperatures and affect conductivity and thermal expansion. The function of Umklapp processes in restricting thermal conductivity is investigated, as well as thermal conductivity and resistivity of the phonon gas, among other thermal transport phenomena. Finally, it is thought that phonon scattering and thermal characteristics are influenced by crystal defects.

REVIEW QUESTIONS

1. Explain the concept of phonons and their significance in the thermal properties of solids. How are phonons different from electrons in terms of their behavior and role in solid-state physics?
2. Compare and contrast the Debye model and the Einstein model for the density of states in solids. What are their respective assumptions and limitations? In what contexts are these models particularly useful?
3. Describe the Debye T^3 law. How does it relate to the low-temperature behavior of the heat capacity of solids? What insights does it provide into the behavior of phonons?
4. Discuss the factors influencing thermal conductivity in solids. How do phonon scattering mechanisms, such as Umklapp processes and defects, affect thermal conductivity? Provide examples of materials where these mechanisms are particularly relevant.
5. Explain the concept of thermal expansion in solids. How is it related to the vibrational modes of the lattice? Discuss how anharmonic interactions influence the thermal expansion coefficient.
6. What role do imperfections and defects play in influencing the thermal properties of solids? How do they affect phonon transport and other thermal phenomena?

MULTIPLE CHOICE QUESTIONS

1. **What concept describes the distribution of energy among phonon modes in a solid?**
   a. Planck distribution
   b. Fermi-Dirac distribution
   c. Bose-Einstein distribution
   d. Maxwell-Boltzmann distribution
2. **Which model approximates the density of states in solids based on a linear relationship between phonon frequency and wave vector?**
   a. Einstein model
   b. Debye model
   c. Fermi model
   d. Planck model
3. **What do Umklapp processes primarily contribute to in the context of phonons?**
   a. Thermal expansion
   b. Phonon-electron interactions
   c. Thermal conductivity
   d. Phonon-phonon interactions
4. **The Debye temperature characterizes the behavior of:**
   a. Electrons in a solid
   b. Phonons in a solid
   c. Neutrons in a solid
   d. Protons in a solid
5. **Which of the following models assumes that phonons in a solid behave like independent quantum particles?**
   a. Classical model
   b. Fermi-Dirac model
   c. Debye model
   d. Einstein model
6. **Anharmonic crystal interactions in solids primarily affect which thermal property?**
   a. Thermal conductivity
   b. Specific heat
   c. Thermal expansion
   d. Phonon frequency

Answers to Multiple Questions

1. (a); 2. (b); 3. (c); 4. (b); 5. (d); 6. (c)

REFERENCES

1. Alexander, S., & Orbach, R. (1982). Density of states on fractals:«fractons». Journal de Physique Lettres, 43(17), 625–631.
2. Allen, P. B., Du, X., Mihaly, L., & Forro, L. (1994). Thermal conductivity of insulating Bi 2 Sr 2 YCu 2 O 8 and superconducting $Bi_2Sr_2CaCu_2O_8$: Failure of the phonon-gas picture. Physical Review B, 49(13), 9073.
3. Anderson, J., & Lapeyre, G. J. (1976). Chemisorption-induced surface umklapp processes in angle-resolved synchrotron photoemission from W (001). Physical Review Letters, 36(7), 376.
4. Askerov, B. M., & Cankurtaran, M. (1994). Isobaric specific heat and thermal expansion of solids in the Debye approximation. Physica status solidi (b), 185(2), 341–348.
5. Blackman, M. (1957). On the thermal expansion of solids. Proceedings of the Physical Society. Section B, 70(9), 827.
6. Buchert, T. (2000). On average properties of inhomogeneous fluids in general relativity: dust cosmologies. General Relativity and Gravitation, 32, 105–125.
7. Chen, C., Lu, T. J., & Fleck, N. A. (1999). Effect of imperfections on the yielding of two-dimensional foams. Journal of the Mechanics and Physics of Solids, 47(11), 2235–2272.
8. Chen, J., Hu, L., Deng, J., & Xing, X. (2015). Negative thermal expansion in functional materials: controllable thermal expansion by chemical modifications. Chemical Society Reviews, 44(11), 3522–3567.
9. Chen, Z., Zhang, X., Lin, S., Chen, L., & Pei, Y. (2018). Rationalizing phonon dispersion for lattice thermal conductivity of solids. National Science Review, 5(6), 888–894.
10. Chernatynskiy, A., Turney, J. E., McGaughey, A. J., Amon, C. H., & Phillpot, S. R. (2011). Phonon-Mediated Thermal Conductivity in Ionic Solids by Lattice Dynamics-Based Methods. Journal of the American Ceramic Society, 94(10), 3523–3531.
11. Chiuri, M. C., & Jappelli, T. (2003). Financial market imperfections and home ownership: A comparative study. European economic review, 47(5), 857–875.
12. Conturo, T. E., McKinstry, R. C., Akbudak, E., & Robinson, B. H. (1996). Encoding of anisotropic diffusion with tetrahedral gradients: a general mathematical diffusion formalism and experimental results. Magnetic resonance in medicine, 35(3), 399–412.
13. Cooper, H. W., & Simmons, G. (1977). The effect of cracks on the thermal expansion of rocks. Earth and Planetary Science Letters, 36(3), 404–412.

14. Cowley, E. R., & Cowley, R. A. (1965). Anharmonic interactions in alkali halides. I. Proceedings of the Royal Society of London. Series A. Mathematical and Physical Sciences, 287(1409), 259–280.
15. Dimitrov, I. K., Manley, M. E., Shapiro, S. M., Yang, J., Zhang, W., Chen, L. D., ... & Li, Q. (2010). Einstein modes in the phonon density of states of the single-filled skutterudite Yb 0.2 Co 4 Sb 12. Physical Review B—Condensed Matter and Materials Physics, 82(17), 174301.
16. Dragunov, T. N., Pavlov, I. S., & Potapov, A. I. (1997). Anharmonic interactions of elastic and orientational waves in one-dimensional crystals. Physics of the Solid State, 39, 118–124.
17. Duke, C. B., & Mahan, G. D. (1965). Phonon-broadened impurity spectra. I. Density of states. Physical Review, 139(6A), A1965.
18. Durczewski, K., Gajek, Z., & Mucha, J. (2020). Influence of electron–phonon interaction and crystal field on thermal and electrical resistivity in rare earth intermetallics. The European Physical Journal B, 93, 1–18.
19. Evans, J. S. O., Mary, T. A., Vogt, T., Subramanian, M. A., & Sleight, A. W. (1996). Negative thermal expansion in ZrW2O8 and HfW2O8. Chemistry of materials, 8(12), 2809–2823.
20. Francl, J., & Kingery, W. D. (1954). Thermal conductivity: IX, experimental investigation of effect of porosity on thermal conductivity. Journal of the American ceramic Society, 37(2), 99–107.
21. Geller, S., Espinosa, G. P., Fullmer, L. D., & Crandall, P. B. (1972). Thermal expansion of some garnets. Materials Research Bulletin, 7(11), 1219–1224.
22. Giamarchi, T. (2004). Theoretical framework for quasi-one dimensional systems. Chemical reviews, 104(11), 5037–5056.
23. Hatke, A. T., Zudov, M. A., Pfeiffer, L. N., & West, K. W. (2011). Phase of phonon-induced resistance oscillations in a high-mobility two-dimensional electron gas. Physical Review B—Condensed Matter and Materials Physics, 84(12), 121301.
24. Hermann, R. P., Jin, R., Schweika, W., Grandjean, F., Mandrus, D., Sales, F. B. C., & Long, G. J. (2003). Einstein oscillators in thallium filled antimony skutterudites. Physical Review Letters, 90(13), 135505.
25. Himpsel, F. J., Altmann, K. N., Bennewitz, R., Crain, J. N., Kirakosian, A., Lin, J. L., & McChesney, J. L. (2001). One-dimensional electronic states at surfaces. Journal of Physics: Condensed Matter, 13(49), 11097.
26. Ivers, D. J., Jackson, A., & Winch, D. (2015). Enumeration, orthogonality and completeness of the incompressible Coriolis modes in a sphere. Journal of Fluid Mechanics, 766, 468–498.
27. Jaffe, G. R., Mei, S., Boyle, C., Kirch, J. D., Savage, D. E., Botez, D., ... & Eriksson, M. A. (2019). Measurements of the thermal resistivity of InAlAs, InGaAs, and InAlAs/InGaAs superlattices. ACS applied materials & interfaces, 11(12), 11970–11975.
28. Jagielski, M., & Kutner, R. (2013). Modelling of income distribution in the European Union with the Fokker–Planck equation. Physica A: Statistical Mechanics and its Applications, 392(9), 2130–2138.

29. Jauernik, S., Hein, P., Gurgel, M., Falke, J., & Bauer, M. (2018). Probing long-range structural order in SnPc/Ag (111) by umklapp process assisted low-energy angle-resolved photoelectron spectroscopy. Physical Review B, 97(12), 125413.

30. Karpov, V. G., & Parshin, D. A. (1985). On the thermal conductivity of glasses at temperatures below the Debye temperature. Zh. Eksp. Teor. Fiz, 88(6), 2212–2227.

31. Kijima, M. (1989). Some results for repairable systems with general repair. Journal of Applied probability, 26(1), 89–102.

32. Kobayashi, K., Ohtsuki, T., Imura, K. I., & Herbut, I. F. (2014). Density of states scaling at the semimetal to metal transition in three dimensional topological insulators. Physical review letters, 112(1), 016402.

33. Lang, L. J., Cai, X., & Chen, S. (2012). Edge states and topological phases in one-dimensional optical superlattices. Physical review letters, 108(22), 220401.

34. Lin, Z., & Zhigilei, L. V. (2007). Temperature dependences of the electron–phonon coupling, electron heat capacity and thermal conductivity in Ni under femtosecond laser irradiation. Applied Surface Science, 253(15), 6295–6300.

35. Lortz, R., Wang, Y., Tutsch, U., Abe, S., Meingast, C., Popovich, P., ... & Junod, A. (2006). Superconductivity mediated by a soft phonon mode: Specific heat, resistivity, thermal expansion, and magnetization of YB 6. Physical Review B—Condensed Matter and Materials Physics, 73(2), 024512.

36. Lu, Y., Sun, T., & Zhang, D. B. (2018). Lattice anharmonicity, phonon dispersion, and thermal conductivity of PbTe studied by the phonon quasiparticle approach. Physical Review B, 97(17), 174304.

37. Lukas, J., Fridrich, J., & Goljan, M. (2005, March). Determining digital image origin using sensor imperfections. In Image and Video Communications and Processing 2005 (Vol. 5685, pp. 249–260).

38. Luther, A., & Peschel, I. (1974). Single-particle states, Kohn anomaly, and pairing fluctuations in one dimension. Physical Review B, 9(7), 2911.

39. Maebashi, H., & Fukuyama, H. (1998). Electrical conductivity of interacting fermions. II. Effects of normal scattering processes in the presence of umklapp scattering processes. Journal of the Physical Society of Japan, 67(1), 242–251.

40. Martin-Delgado, M. A. (1991). Planck distribution for a q-boson gas. Journal of Physics A: Mathematical and General, 24(21), L1285.

41. Maznev, A. A., & Wright, O. B. (2014). Demystifying umklapp vs normal scattering in lattice thermal conductivity. American journal of physics, 82(11), 1062–1066.

42. Milman, V. Y., Grishchenko, T. A., Evlashina, M. L., Zhalko-Titarenko, A. V., Antonov, V. N., & Nemoshkalenko, V. V. (1992). Theoretical study of lattice heat capacity, thermal expansion and phonon-limited resistivity of transition metals. Journal of Physics and Chemistry of Solids, 53(4), 565–575.

43. Moon, M. W., Chung, J. W., Lee, K. R., Oh, K. H., Wang, R., & Evans, A. G. (2002). An experimental study of the influence of imperfections on the buckling of compressed thin films. Acta Materialia, 50(5), 1219–1227.

44. Naumov, V. N., & Musikhin, A. E. (2015). Numerical solution of the inverse problem

of reconstructing phonon density of states from experimental heat capacity. Physica B: Condensed Matter, 476, 41–49.

45. Nipko, J. C., & Loong, C. K. (1998). Phonon excitations and related thermal properties of aluminum nitride. Physical Review B, 57(17), 10550.
46. Padture, N. P., & Klemens, P. G. (1997). Low thermal conductivity in garnets. Journal of the American Ceramic Society, 80(4), 1018–1020.
47. Pannocchia, G., Rawlings, J. B., & Wright, S. J. (2007). Fast, large-scale model predictive control by partial enumeration. Automatica, 43(5), 852–860.
48. Rosenberg, H. M. (1955). The thermal conductivity of metals at low temperatures. Philosophical Transactions of the Royal Society of London. Series A, Mathematical and Physical Sciences, 247(933), 441–497.
49. Sanati, M., Albers, R. C., Lookman, T., & Saxena, A. (2011). Elastic constants, phonon density of states, and thermal properties of UO 2. Physical Review B—Condensed Matter and Materials Physics, 84(1), 014116.
50. Schafer, B. W., & Peköz, T. (1998). Computational modeling of cold-formed steel: characterizing geometric imperfections and residual stresses. Journal of constructional steel research, 47(3), 193–210.
51. Schapery, R. A. (1968). Thermal expansion coefficients of composite materials based on energy principles. Journal of Composite materials, 2(3), 380–404.
52. Schliesser, J. M., & Woodfield, B. F. (2015). Development of a Debye heat capacity model for vibrational modes with a gap in the density of states. Journal of Physics: Condensed Matter, 27(28), 285402.
53. Slack, G. A., Tanzilli, R. A., Pohl, R. O., & Vandersande, J. W. (1987). The intrinsic thermal conductivity of AIN. Journal of Physics and Chemistry of Solids, 48(7), 641–647.
54. Smith, D. S., Alzina, A., Bourret, J., Nait-Ali, B., Pennec, F., Tessier-Doyen, N., ... & Gonzenbach, U. T. (2013). Thermal conductivity of porous materials. Journal of Materials Research, 28(17), 2260–2272.
55. Stapleton, R. C., & Subrahmanyam, M. G. (1977). Market imperfections, capital market equilibrium and corporation finance. The Journal of Finance, 32(2), 307–319.
56. Sussmann, J. A., & Thellung, A. (1963). Thermal conductivity of perfect dielectric crystals in the absence of umklapp processes. Proceedings of the Physical Society, 81(6), 1122.
57. Thoss, M., Wang, H., & Miller, W. H. (2001). Self-consistent hybrid approach for complex systems: Application to the spin-boson model with Debye spectral density. The Journal of Chemical Physics, 115(7), 2991–3005.
58. Vali, R. (2008). Phonons and heat capacity of LaAlO3. Computational materials science, 44(2), 779–782.
59. Wagner, L. O., Stoudenmire, E. M., Burke, K., & White, S. R. (2012). Reference electronic structure calculations in one dimension. Physical Chemistry Chemical Physics, 14(24), 8581–8590.

60. Wang, Y., Shang, S. L., Fang, H., Liu, Z. K., & Chen, L. Q. (2016). First-principles calculations of lattice dynamics and thermal properties of polar solids. npj Computational Materials, 2(1), 1–10.

61. Westphal, D., & Goldmann, A. (1983). Adsorbate-induced umklapp processes in photoemission from Cl on Cu. Surface Science, 126(1–3), 253–257.

62. Williams, R. K., Butler, W. H., Graves, R. S., & Moore, J. P. (1983). Experimental and theoretical evaluation of the phonon thermal conductivity of niobium at intermediate temperatures. Physical Review B, 28(11), 6316.

63. Wilson, A. C. (1942). Imperfections in the structure of cobalt II. Mathematical treatment of proposed structure. Proceedings of the Royal Society of London. Series A. Mathematical and Physical Sciences, 180(982), 277–285.

64. Xiao-Hong, S., Jiu-Xun, S., & Chun-Hua, X. (2016). Dependence of general Einstein relation on density of state for organic semiconductors. Organic Electronics, 35, 65–73.

65. Xu, S. M., & Bogers, M. L. (2024). Imperfections-as-practice: Projects as becoming processes of imperfections. Project Management Journal, 55(2), 151–166.

66. Young, J., Stroppa, A., Picozzi, S., & Rondinelli, J. M. (2015). Anharmonic lattice interactions in improper ferroelectrics for multiferroic design. Journal of Physics: Condensed Matter, 27(28), 283202.

67. Zhang, Q., Chen, G., Yoon, S. F., Ahn, J., Wang, S. G., Zhou, Q., ... & Li, J. Q. (2002). Thermal conductivity of multiwalled carbon nanotubes. Physical Review B, 66(16), 165440.

68. Zudov, M. A., Ponomarev, I. V., Efros, A. L., Du, R. R., Simmons, J. A., & Reno, J. L. (2001). New class of magnetoresistance oscillations: Interaction of a two-dimensional electron gas with leaky interface phonons. Physical review letters, 86(16), 3614.

CHAPTER

5

Absorption of Light in Solids

LEARNING OBJECTIVES

After studying this chapter, you will be able to:

- Understand the concept of the absorption coefficient and its role in optical measurements in solids;
- Understand the frequency-dependent absorption characteristics of free carriers in semiconductors;
- Understand free carrier absorption characteristics in metals under specific frequency conditions;
- Understand direct interband transitions' absorption coefficient and its dependencies;
- Understand conservation of crystal momentum in direct optical transitions;
- Understand energy and momentum conservation in indirect interband transitions.

KEY TERMS FROM THIS CHAPTER

Absorption coefficient
Complex index of refraction
Extinction coefficient
Indirect interband transitions
Poynting vector
Complex dielectric function
Direct interband transitions
Free carrier absorption
Metals
Semiconductors

UNIT INTRODUCTION

The absorption of light in solids is a fundamental and extensively studied phenomenon in condensed matter physics, crucial for understanding the interaction between photons and electrons within solid materials. When a photon encounters a solid, it can transfer its energy to an electron, causing the electron to transition from a lower energy state (typically within the valence band) to a higher energy state (in the conduction band). This process results in the creation of an electron-hole pair, where the hole represents the absence of an electron in the valence band.

The probability and efficiency of this energy transfer depend on the energy of the incident photon, which is directly related to its frequency and thus its wavelength. This relationship is quantified by Planck's constant . Each solid material possesses a unique absorption spectrum, dictating the wavelengths of light it can effectively absorb. This spectrum is intimately tied to the band structure of the material, which describes the distribution of electronic energy levels across its atomic lattice.

The optical properties of solids, such as their color, transparency, refractive index, reflectivity, and transmission characteristics, are directly influenced by their absorption behavior. Materials that selectively absorb certain wavelengths of visible light appear colored, while those that absorb minimally in the visible range can appear transparent. Understanding and manipulating these absorption properties are pivotal for various technological applications.

The chapter explores fundamental principles governing how materials interact with light. Topics include the absorption coefficient, which quantifies light absorption efficiency, and mechanisms such as free carrier absorption in semiconductors and metals. The chapter also covers direct and indirect interband transitions, addressing their impact on optical properties. Temperature effects on band gap energy and dependencies of absorption edge on Fermi energy and applied electric fields are discussed, along with principles of crystal momentum conservation in optical transitions.

5.1 THE ABSORPTION COEFFICIENT

For optical measurements in materials, it is essential to comprehend light absorption. During absorption measurements, we analyze the incident intensity in relation to the light intensity I(z) after it passes through a specific thickness z of material (Mott, 1938).

This allows us to calculate the absorption coefficient, or (ω):

$$I(z) = I_0 e^{-\alpha_{abs}(w)z} \tag{5.1}$$

Figure 5.1 provides a schematic representation of the absorption constant. It may be inferred from the relationship between the square of the field variables and the intensity I(z) that:

$$\alpha_{abs}(w) = 2\frac{w\tilde{k}(w)}{c} \tag{5.2}$$

The definition of (ω) for light intensity, which is precisely proportional to the square of the fields, provides the reason for the factor of 2. The absorption coefficient and the imaginary component of the complex index of refraction, or extinction coefficient, also known as (ω), are directly related, according to this equation (Thomasson, 1980).

As a result, k˜ is often associated with power outages. It is crucial to remember that, provided the conditions and are met, Equation 5.2 relates to the occurrence of free carrier absorption in semiconductors.

It is clear that distinct physical processes influencing the optical characteristics of solids result in notably diverse frequency dependences of the absorption coefficient (Wang et al., 2015). The discussion that follows will examine the absorption coefficient's frequency dependency for a particular subject.

Exploring the phenomenon of free carrier absorption:

- The typical semiconductor has an absorption coefficient (ω) that is inversely proportional to the square of the angular frequency, ;
- Low frequencies metals (ω) ~ .

Interband transitions that occur directly (Pace et al., 1995):

- Form of absorption coefficient (ω)~;
- Crystal momentum conservation;
- Relation between and momentum matrix element;
- Form of (ω) for direct forbidden transition ~.

Indirect interband transitions (Lin et al., 1973):

- Form of absorption coefficient (ω)~.

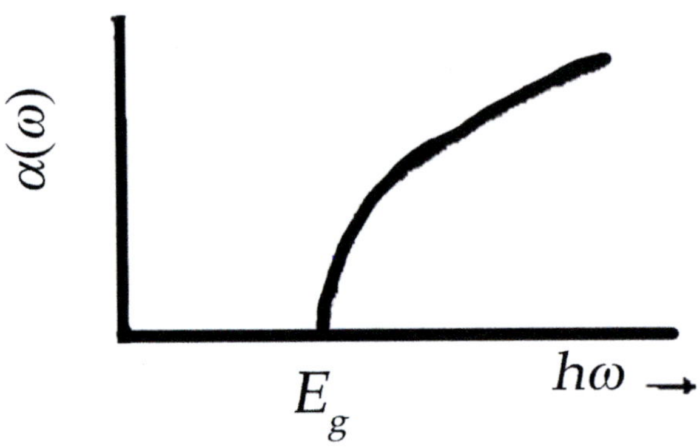

Figure 5.1. Illustration of absorption coefficient frequency dependance close to an interband transition threshold.

Source: M. S. Dresselhaus, Creative Commons License.

The previous overview relates to three-dimensional systems. The dimensionality of the system in each process influences the interaction between functions in one-dimensional and two-dimensional systems (Rajkanan et al., 1979).

5.2. FREE CARRIER ABSORPTION IN SEMICONDUCTORS

To find out the free carrier absorption, the complex dielectric function equation ε(ω) = (ω) + (ω) can be used:

$$\varepsilon(w) = \varepsilon_0 + \frac{4\pi i\sigma}{w} \quad (5.3)$$

The symbol represents the dielectric constant of the core material at optical frequencies above the lattice mode frequencies. In this context, is assumed to be unaffected by the angular frequency ω. The Drude term determines the link between electronic polarizability and frequency-dependent electrical conductivity (Zhang et al., 2022).

$$\sigma = \frac{ne^2\tau}{m^*(1-iw\tau)} \quad (5.4)$$

The plasma frequency is determined by the point at which (ω), becomes zero, specifically () = 0 or:

$$w_p^2 = \frac{4\pi ne^2}{m^*\varepsilon_0} \quad (5.5)$$

In semiconductors, the ε_0 (dielectric constant) has generally a significant value, with the impact of free carriers being relatively minimal when it comes to infrared and visible frequencies. In the context of metals, the absorption of free carriers reigns supreme across the entire spectrum of optical frequencies (Schroder et al., 1978).

In the case of semiconductors, the frequency range that is typically considered is higher than the frequencies associated with optical phonons. In this range, it is generally observed that the product of the angular frequency (ω) and the relaxation time (τ) is significantly more than 1. Subsequently, we may compose:

$$\varepsilon(\omega) = \varepsilon_0 + \frac{4\pi i n e^2 \tau(1+i\omega\tau)}{m^*\omega[1+(\omega\tau)^2]} = \varepsilon_0 + \frac{i\varepsilon_0\omega_p^2\tau(1+i\omega\tau)}{\omega[1+(\omega\tau)^2]} \tag{5.6}$$

or for :

$$\varepsilon(\omega) \simeq \varepsilon_0 + \frac{i\varepsilon_0\omega_p^2\tau^2}{\omega^3\tau^3} - \frac{\varepsilon_0\omega_p^2}{\omega^2}. \tag{5.7}$$

For optical measurements in semiconductors, it is important to note that the relation is satisfied within the range of interest. One way to simplify the expression of the complex dielectric function ε (ω) is by using the optical constants n˜(ω) and k˜(ω). This can be done through the definition . The exponent of 2 in this context, the index of refraction is denoted as (ω) and the extinction coefficient is represented as k˜(ω). Next, the real part of the dielectric function can be written as (Schumann & Phillips, 1967):

$$\varepsilon_1(\omega) \equiv \tilde{n}^2(\omega) - \tilde{k}^2(\omega) \approx \varepsilon_0 \tag{5.8}$$

where: the refractive index n˜(ω) has a high value and the extinction coefficient k ˜(ω) has a low value. The dielectric function›s imaginary component is given by:

$$\varepsilon_2(\omega) \equiv 2\tilde{n}(\omega)\tilde{k}(\omega) \approx 2\sqrt{\varepsilon_0}\,\tilde{k}(\omega) = \frac{\varepsilon_0\omega_p^2\tau^2}{\omega^3\tau^3} \tag{5.9}$$

which is relatively small, since ω. Hence, the coefficient of the absorption is:

$$\alpha_{\text{abs}}(\omega) = \frac{2\omega\tilde{k}(\omega)}{c} \simeq \frac{2\omega}{c}\frac{\varepsilon_0\omega_p^2}{2\sqrt{\varepsilon_0}\omega^3\tau} = \frac{\sqrt{\varepsilon_0}\omega_p^2}{c\omega^2\tau} \tag{5.10}$$

Therefore, (ω) is directly proportional to 1/ or to when taking into account semiconductor's free carrier absorption under the conditions where ωτ ≫1 and ω ≫ . Figure 5.2 displays a graph that represents the relationship between the optical coefficient of absorption of InAs and the wavelength. The graph is presented on a logarithmic scale. It demonstrates that the coefficient of absorption, (ω), is proportional to the wavelength raised to the power of p. The value of p lies within the range of 2 to 3 over a wide spectrum of donor concentrations (James & Smith, 1982). Figure 5.3 illustrates the relationship between the reflectivity spectra and the wavelength for different concentrations of donors in severely doped n-type InSb. These results clearly show the link between the carrier concentration and the plasma frequency (Bonod et al., 2008).

5.3. METALS AND FREE CARRIER ABSORPTION

There are certain variations in the limits between metals and semiconductors. Here, we focus on the scenario where ωτ 1, ω , 4πσ/ω, resulting in an approximate equality (Jain et al., 1991) between n˜ and k˜. Thus, we obtain:

$$\varepsilon(\omega) \simeq \frac{4\pi i\sigma}{\omega} \simeq \frac{4\pi ine^2\tau}{\omega m^*} \simeq i\varepsilon_2(\omega) \equiv 2i\tilde{n}\tilde{k} \simeq 2i\tilde{k}^2 \tag{5.11}$$

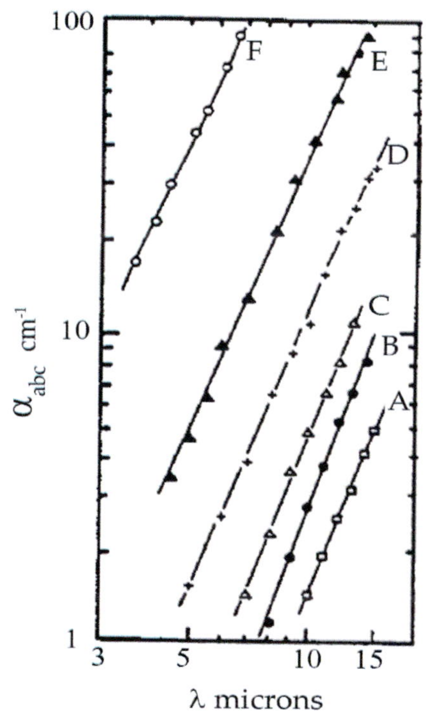

Figure 5.2. Illustration of free carrier absorption for six distinct carrier concentrations (in units of) in n-type In As at room temperature, A: 0.28; C: 1.4; B: 0.85; E: 7.8; D: 2.5; and F: 39.0.

Source: Ziman, Creative Commons License.

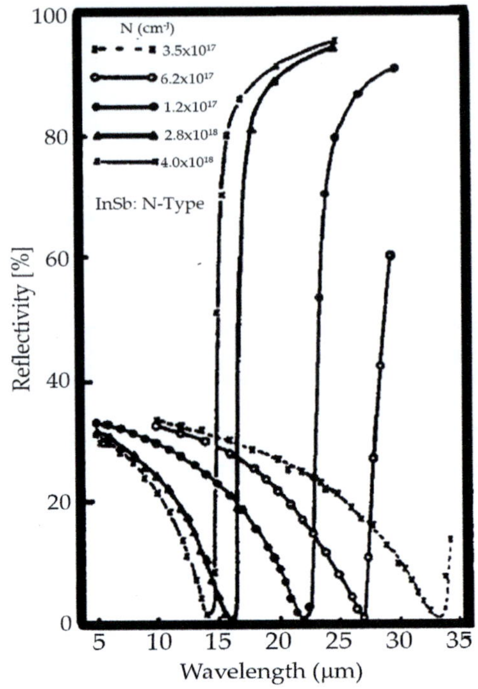

Figure 5.3. Illustration of the reflectivity spectra of n-type InSb at ambient temperature show the presence of plasma edges, with carrier concentration n ranging from 3.5 × and 4.0×. The solid curves are theoretical models that accurately incorporate the energy dependence of strong interband coupling, commonly known as non-parabolic effects, in relation to the effective mass (m*) of the system. This provides us with (ω) for the extinction coefficient.

Source: Bassani, Creative Commons License.

This provides us with k˜ (ω) for the extinction coefficient:

$$\tilde{k}(\omega) = \sqrt{\frac{2\pi n e^2 \tau}{m^* \omega}} \tag{5.12}$$

where the coefficient of absorption is:

$$\alpha_{\text{abs}}(\omega) = \frac{2\omega \tilde{k}(\omega)}{c} = \sqrt{\frac{8\pi \omega n e^2 \tau}{m^* c^2}} \tag{5.13}$$

In this particular limit, the value of (ω) is directly related to the square root of . Typically, the reflectivity is the preferred observable for metals. When considering metals, it is important to note that, and both n˜ and k˜ are large in the appropriate limit (Schroder et al., 1978). Therefore, we have:

$$\mathcal{R} = \frac{(\tilde{n}-1)^2 + \tilde{k}^2}{(\tilde{n}+1)^2 + \tilde{k}^2} = \frac{\tilde{n}^2 - 2\tilde{n} + 1 + \tilde{k}^2}{\tilde{n}^2 + 2\tilde{n} + 1 + \tilde{k}^2} = 1 - \frac{4\tilde{n}}{\tilde{n}^2 + \tilde{k}^2 + 2\tilde{n} + 1} \tag{5.14}$$

$$\mathcal{R} \approx 1 - \frac{4\tilde{n}}{\tilde{n}^2 + \tilde{k}^2} \approx 1 - \frac{2}{\tilde{n}}. \tag{5.15}$$

However, by utilizing Equation 5.12 and considering the condition n˜=k˜ ≫1, one can derive:

$$\tilde{n}(\omega) \simeq \sqrt{\frac{2\pi m e^2 \tau}{m^* \omega}} \tag{5.16}$$

in order for the reflectivity to increase accordingly:

$$\mathcal{R}(\omega) \simeq 1 - 2\sqrt{\frac{m^* \omega}{2\pi n e^2 \tau}}. \tag{5.17}$$

Equation 5.17, commonly referred to as the Hagen-Rubens relation, is widely applicable to several metals in infrared range of electromagnetic spectrum. It can also be used for degenerate semiconductors that operate at frequencies below the plasma frequency (Linnros, 1998).

5.4. DIRECT INTERBAND TRANSITIONS

To determine absorption from direct interband transitions, one can use the absorption coefficient (ω). This coefficient measures the power absorbed by a substance per unit volume and per unit electromagnetic energy incoming flux (Hordvik & Schlossberg, 1977).

$$\alpha_{\text{abs}}(\omega) = \frac{(\hbar\omega) \times \text{number of transitions/unit volume/unit time}}{\text{incident electromagnetic flux}}. \tag{5.18}$$

Calculating the incident electromagnetic flux involves determining the Poynting vector.

$$\vec{S} = \frac{c}{8\pi} Re(\vec{E}^* \times \vec{H}). \tag{5.19}$$

It is useful to establish a connection between the field variables and the vector potential:

$$\vec{E} = -\frac{1}{c}\frac{\partial \vec{A}}{\partial t} = \frac{i\omega}{c}\vec{A} \tag{5.20}$$

$$\mu\vec{H} = \vec{B} = \vec{\nabla} \times \vec{A}. \tag{5.21}$$

Permeability μ can be considered as unity in non-magnetic materials. When considering the curl of , it is assumed to have a plane wave form (Koffyberg et al., 1979).

$$\vec{A} = \vec{A}_0 e^{i(\vec{K}\cdot\vec{r} - \omega t)} \tag{5.22}$$

The wave vector represents the propagation constant for the light. Therefore, we derive the expression for the Poynting vector.

$$\vec{S} = \frac{c}{8\pi} Re\left[-\frac{i\omega}{c} \vec{A}^* \times (i\vec{K} \times \vec{A}) \right] \tag{5.23}$$

$$\vec{S} = \frac{\omega}{8\pi} Re\left[(\vec{A}^* \cdot \vec{A})\vec{K} - (\vec{A}^* \cdot \vec{K})\vec{A} \right]. \tag{5.24}$$

Utilizing the fact that ∗. = 0 for a transverse plane wave:

$$\vec{S} = \frac{\omega}{8\pi}\frac{\tilde{n}\omega}{c}|A|^2\hat{K} \tag{5.25}$$

where: the unit vector is aligned with the Poynting vector; and the symbol denotes the real part of the complex index of refraction (Denninger et al., 1982). The absorption coefficient is defined by Equation 5.18, where the denominator is the quantity from Equation 5.25. The widely recognized "Fermi Golden Rule" is used for calculating the chance of a transition occurring per unit time per unit volume (Garriga et al., 1987).

$$W = \frac{2\pi}{\hbar} \mid \mathcal{H}'_{vc} \mid^2 \rho_{cv}(\hbar\omega). \tag{5.26}$$

In order to properly characterize the occupation of the states at a given temperature, the Fermi functions must be included when examining the absorption process at a non-zero temperature.

$$f(E_v)[1 - f(E_c)] - f(E_c)[1 - f(E_v)] \tag{5.27}$$

in which the first group of terms represents the absorption process which depends on the valence band (v) being nearly full and the conduction band (c) being nearly empty. The second group of terms represents the emission process which proceeds if there are occupied conduction states and unoccupied valence states (Mitra & Lambrecht, 2008).

It is clear that . is the simplified form of the Fermi functions found in Equation 5.27. Equation 5.26 use the electromagnetic interaction Hamiltonian to express the matrix elements .

$$\mathcal{H}'_{vc} = \langle v|\mathcal{H}'_{em}|c\rangle = -\left(\frac{e}{mc}\right)\langle v|\vec{A}(\vec{r},t)\cdot\vec{p}|c\rangle. \tag{5.28}$$

In section 5.5, we demonstrate that the diagonal matrix element, represents the electromagnetic interaction between valence and conduction bands in wave vector . This is because the light wave vector, is much smaller than the Brillouin zone dimensions. Therefore, it is also possible to ignore the spatial variation of the vector potential. Therefore, the square of the matrix elements connecting the conduction and valence bands can be derived (Yamaguchi & Nobusada, 2016).

$$|\mathcal{H}'_{vc}|^2 = \left(\frac{e}{mc}\right)^2 |A|^2 |\langle v|p|c\rangle|^2, \tag{5.29}$$

Where in the conduction and valence bands, there is a coupling of states with the same electron wave vector, denoted as |. In contrast to , varies slowly with k. Therefore, it is convenient to ignore the k dependency of this quantity and assess it at the critical point (Segura et al., 2021). The absorption coefficient of direct interband transitions can be derived as follows:

$$\alpha_{\text{abs}}(\omega) = \frac{(\hbar\omega)[\frac{2\pi}{\hbar}(\frac{e}{mc})^2|A|^2|\langle v|p|c\rangle|^2\rho_{cv}(\hbar\omega)][f(E_v) - f(E_c)]}{\frac{\omega}{8\pi}\frac{\tilde{n}\omega}{c}|A|^2} \quad (5.30)$$

$$\alpha_{\text{abs}}(\omega) = \frac{16\pi^2 e^2}{m^2 c\tilde{n}\omega}|\langle v|p|c\rangle|^2\rho_{cv}(\hbar\omega)[f(E_v) - f(E_c)] \quad (5.31)$$

where: the index of refraction is indicated by in equations 5.30 and 5.31.

To understand the functional forms of the values in Equation 5.31, let us consider a simple case with two parabolic bands. These bands show an allowed optical transition, i.e., they are coupled by a non-zero momentum matrix element (Malyi & Acosta, 2020) describing the joint density of states for a critical point (found in many semiconductors, often around k = 0);

$$\rho_{cv}(\hbar\omega) = \frac{1}{2\pi^2}\left(\frac{2m_r}{\hbar^2}\right)^{3/2}\sqrt{\hbar\omega - E_g} \quad (5.32)$$

By reducing the mass of the valence and conduction bands, we can estimate the absorption coefficient (ω) (Lahbabi et al., 2000). In the field of low temperatures, semiconductors exhibit a valence band that is predominantly occupied and a conduction band that remains vacant; specifically, = 1 and = 1. From the effective mass sum-rule, we can calculate the square of

$$|\langle v|p|c\rangle|^2 \simeq \frac{m_0 E_g}{2}\frac{m_0}{m^*} \quad (5.33)$$

where: represents the mass of a free electron. Equations 5.33 and 5.32 can be substituted into Equation 5.31 to drive the absorption coefficient frequency dependence in direct allowed transitions (Lambrecht, 2000).

$$\alpha_{\text{abs}}(\omega) \propto \frac{1}{\omega}\sqrt{\hbar\omega - E_g} \quad (5.34)$$

Figure 5.1 shows that the interband transitions that are directly permitted by optics have a minimum energy need at the energy gap E_g. For example. The frequency dependence of (ω) varies significantly across different physical processes (Dou et al., 2014).

Occasionally, it can be beneficial to express the absorption by utilizing the imaginary component of the dielectric function.

$$\varepsilon_2(\omega) = \frac{\tilde{n}c}{\omega}\alpha_{\text{abs}}(\omega) \quad (5.35)$$

which from Equation 5.31 becomes:

$$\varepsilon_2(\omega) = \left(\frac{4\pi e}{m\omega}\right)^2 |\langle v|p|c\rangle|^2 \rho_{cv}(\hbar\omega)[f(E_v) - f(E_c)]. \tag{5.36}$$

Let us define the dimensionless quantity , commonly referred to as the oscillator strength, as:

$$f_{vc} = \frac{2|\langle v|p|c\rangle|^2}{m[E_c(k) - E_v(k)]} = \frac{2|\langle v|p|c\rangle|^2}{m\hbar\omega}, \tag{5.37}$$

The result for (ω) at T = 0 is as follows:

$$\varepsilon_2(\omega) = \left(\frac{8\pi^2 e^2 \hbar}{m\omega}\right) f_{vc}\rho_{cv}(\hbar\omega). \tag{5.38}$$

The Kramers-Kronig relation is used to derive the value of (ω) for interband transitions based on the value of (ω).

A graph showing the relationship between the basic models and the absorption coefficient's behavior at the primary absorption threshold is shown in Figure 5.4.

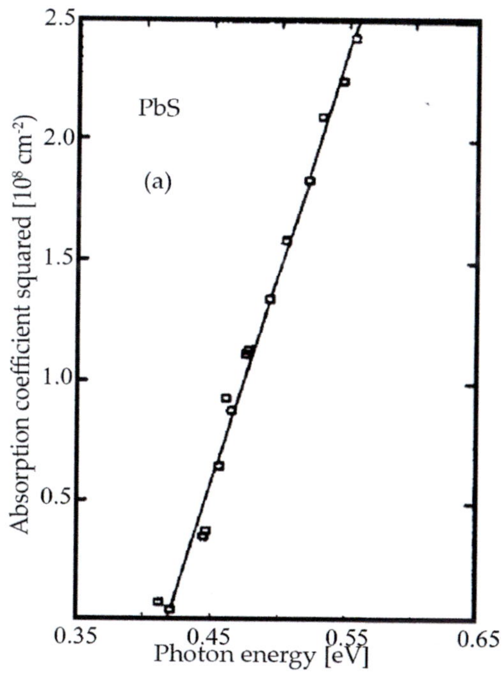

Figure 5.4. The graph shows the relationship between the absorption coefficient of PbS square and the energy of photons, demonstrating a linear dependency on hω. The point where the line intersects the x-axis determines the direct energy gap.

Source: Cardona, Creative Commons License.

The relationship between versus for PbS is displayed in this plot. To determine the direct energy band gap, one can examine the point where intersects with the photon energy axis. Plotting (ω) against on a logarithmic scale reveals the energy gap, as demonstrated in Figure 5.5 for InSb (Guliamov et al., 2001).

The method used to obtain Equation 5.31 is also utilized to derive the functional expression for the absorption coefficient in the case of direct prohibited transitions. However, in this case, the square of the magnitude of the product of the velocity (Lyu et al., 2023), momentum, and speed of light is now dependent .

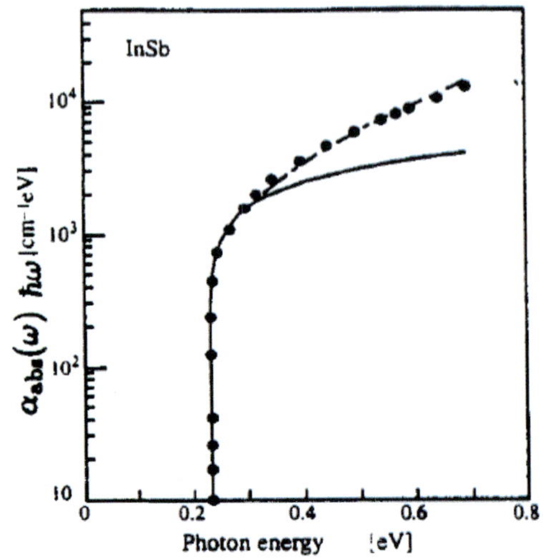

Figure 5.5. Plotting the InSb absorption coefficient semilogarithmically against photon energy at 5K. The filled circles show the outcomes of the experiments. A variety of models have been used to calculate the curves. The inclusion of the matrix elements reliance on k yields the best results. The direct bandgap of InSb is given by the intercept with the x-axis.

Source: Wolfe, Creative Commons License.

5.4.1. Temperature Dependence of

The different band characteristics, especially the energy gap, are temperature dependent due to the lattice's expansion and contraction with temperature. Although there are algorithms that can predict and take into consideration the band gap temperature dependence of at the fundamental absorption edge (threshold), empirical fits are still the best method for determining (Šćepanović et al., 2007).

Below are some relevant expressions for these fits that can be used in research.

$E_g(T) = 1.165 - 2.84 \times 10^{-4}T$ (eV) Si

$E_g(T) = 0.742 - 3.90 \times 10^{-4}T$ (eV) Ge

$E_g(T) = 1.522 - \frac{5.8\times10^{-4}T^2}{T+300}$ (eV) GaAs

$E_g(T) = 2.338 - \frac{6.2\times10^{-4}T^2}{T+460}$ (eV) GaP

As previously demonstrated, declines with rising T for semiconductors in Group IV and III–V compounds (Schwendel et al., 2001), but increases with increasing T for compounds in Group IV–VI.

5.4.2. The Absorption Edges Dependency on Fermi Energy

Neglecting the excitonic effects that are addressed, lies in the bandgap for weakly doped semiconductors, and the absorption edge occurs at .

On the other hand, the threshold for optical absorption is moved in strongly doped semiconductors, where is located in the conduction or valence bands (Jasieniak et al., 2011).

The Burstein shift, also known as the movement in the absorption edge, is shown in Figure 5.6. It can be seen that the absorption threshold occurs when:

$$\hbar\omega = E_g + \frac{\hbar^2 k_{\rm BS}^2}{2}\left(\frac{1}{m_e^*} + \frac{1}{m_h^*}\right) = E_g + \frac{\hbar^2 k_{\rm BS}^2}{2m_r^*} = E_F - 4k_B T$$

(5.39)

where: the reduced mass is denoted by , the wave vector corresponds to Burstein shift given in Equation 5.39, and ((1/) = (1/) + (1/).

When it comes to heavily doped semiconductors, it is worth noting that the doping process causes a shift in the position of the Fermi level. Consequently, the behavior of the Fermi functions is dependent on the concentration of carriers (as shown in Equation 5.27), taking into account the possibility of the initial state being occupied and the final state being unoccupied. The Burstein shift will specifically alter the amount (1 –) indicating the availability of final states (Merino et al., 2017). In the event that we write

$$\frac{\hbar^2 k_{\rm BS}^2}{2m_e^*} = E - E_c$$

(5.40)

Consider , the energy at the bottom of the conduction band, to calculate the probability of an empty ultimate state (Wang et al., 2022).

$$1 - f_0 = \frac{1}{1 + \exp[(E_F - E)/k_B T]} = \frac{1}{1 + \exp\left[\frac{E_F - E}{k_B T} - \frac{(\hbar\omega - E_g)m_h^*}{(m_e^* + m_h^*)k_B T}\right]}$$

(5.41)

When evaluating in Equation 5.31, it is necessary to utilize Equation 5.41 for calculating the probability of final states (Katahara & Hillhouse, 2014).

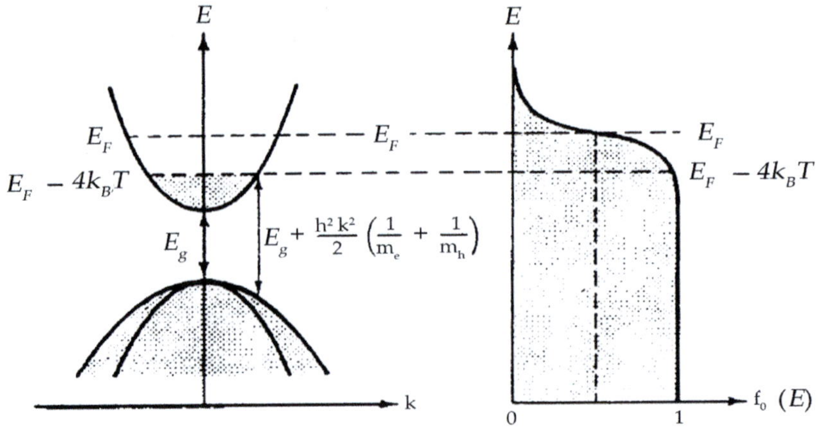

Figure 5.6. Illustration showing doping of an n-type semiconductor causes a shift of the basic absorption edge towards higher energy levels. The Burstein shift wave vector, denoted as kBS is represented in equation 5.39.

Source: Stillman, Creative Commons License.

5.4.3. Dependence of Absorption Edge on Applied Electric Field

Within the energy gap, the valence and conduction bands' electron wave functions show an exponential amplitude reduction. When an electric field is present, an electron in the valence band must undergo tunneling through a triangle barrier in order to get to the conduction band (Aspnes, 1966).

As shown in Figure 5.7(a), the barrier's height and thickness when there is no photon absorption are and respectively, where is an electric field magnitude (Miller et al., 1985).

The photon, as shown in Figure 5.7(b), reduces the thickness of the barrier.

$$t(\hbar\omega) = \frac{E_g - \hbar\omega}{e|\vec{E}|}$$

(5.42)

In order to increase the probability of tunneling through the absorption of photons. The absorption edge is greatly diminished by the electric field, as Figure 5.8 shows, and below the zero-field band gap, its effect on is most evident. The phenomenon where an electric field changes a material's absorption edge is known as the Franz-Keldysh effect (Miller et al., 1986).

5.5. CONSERVATION OF CRYSTAL MOMENTUM IN DIRECT OPTICAL TRANSITIONS

Let us explain why momentum matrix elements between two Bloch states in a perfect crystal are diagonal and how they preserve crystal momentum for better visibility. The momentum matrix elements have a unique characteristic that permits direct interband transfers (Vasquez, 1998). We

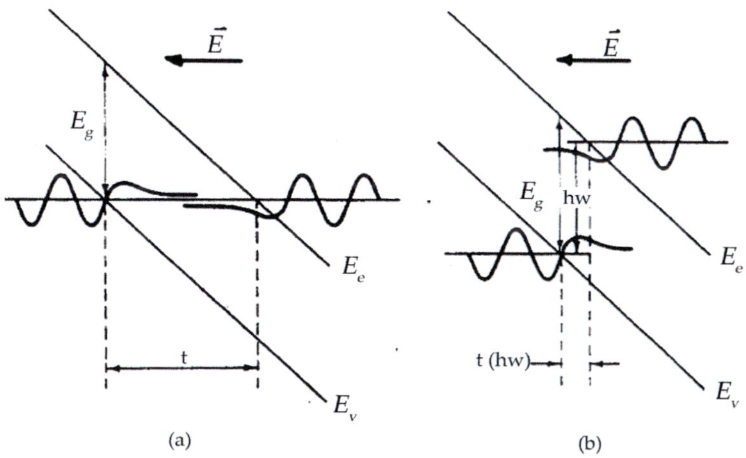

Figure 5.7. Diagram of energy band in an electric field displaying the overlap of the wavefunction (a) in the absence of photon absorption (b) in the presence of such absorption.

Source: Cardona, Creative Commons License.

The elements of the momentum matrix that couple two bands, like the conduction and valence bands, are as follows:

$$\langle n'\vec{k}'|\vec{p}|n,\vec{k}\rangle = \int d^3r e^{-i\vec{k}'\cdot\vec{r}} u^*_{n'k'}(\vec{r}) \left(\frac{\hbar}{i}\vec{\nabla}\right) e^{i\vec{k}\cdot\vec{r}} u_{nk}(\vec{r}). \tag{4.43}$$

Using on the Bloch state's product function produces:

$$\langle n'\vec{k}'|\vec{p}|n,\vec{k}\rangle = \int d^3r e^{-i\vec{k}'\cdot\vec{r}} u^*_{n'k'}(\vec{r}) e^{i\vec{k}\cdot\vec{r}} (\hbar\vec{k} + \frac{\hbar}{i}\vec{\nabla}) u_{nk}(\vec{r}). \tag{4.44}$$

The term in h is diagonal in the band index and crystal momentum, as shown by the simple integration that yields . Interband transitions are hence not produced by this phrase (Gobeli et al., 1964).

The term that is left in Equation 5.44 is:

$$\int d^3r e^{i(\vec{k}-\vec{k}')\cdot\vec{r}} \frac{\hbar}{i} u^*_{n'k'}(\vec{r}) \vec{\nabla} u_{nk}(\vec{r}). \tag{5.45}$$

Under the translation → → → + n, where n is any lattice vector, the function in Equation 5.45 is periodic. However, Fourier expansion can be applied to any spatially periodic function (Behzad, 2019).

$$\sum_m F_m e^{i\vec{G}_m\cdot\vec{r}} = \frac{\hbar}{i} u^*_{n'k'}(\vec{r}) \nabla u_{nk}(\vec{r}) \tag{5.46}$$

Using the reciprocal lattice vectors m as a basis. Consequently, we derive for the integral in Equation 5.45.

$$\int d^3r e^{i(\vec{k}-\vec{k}')\cdot\vec{r}} F_m e^{i\vec{G}_m\cdot\vec{r}} \tag{5.47}$$

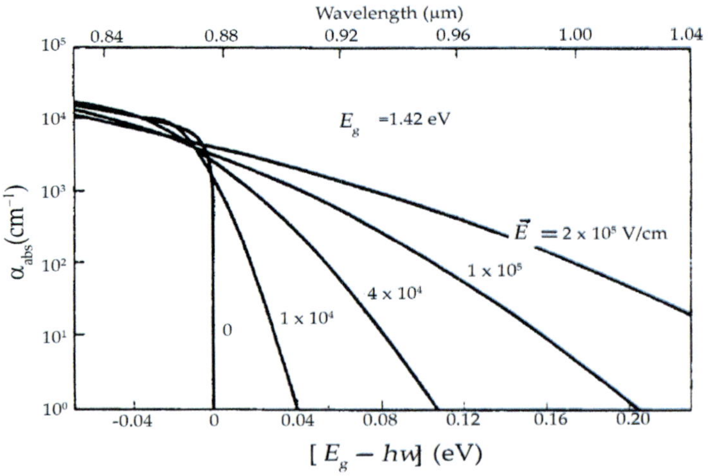

Figure 5.8. Illustration of GaAs band to band absorption that depends on photon energy and electric field.

Source: Pastori Parravicini, Creative Commons License.

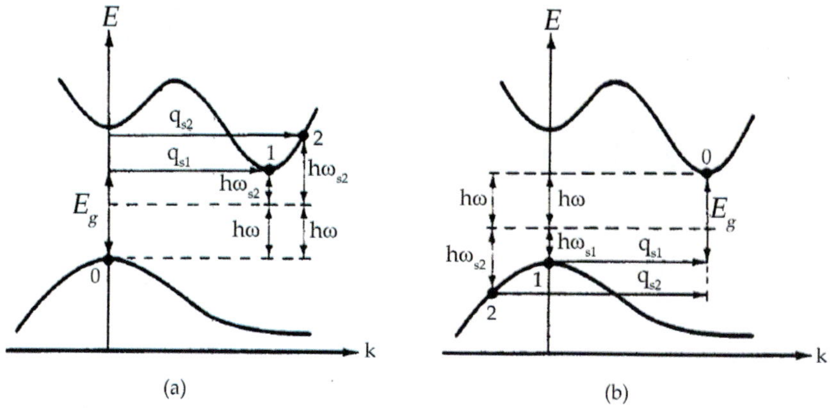

Figure 5.9. The electrons undergo indirect optically induced (a) they transition in the valence bond from initial state 0 to final states 1 and 2 b) the transition from the initial states 1 and 2 in valence bond to final state 0 in conduction band.

Source: M. S. Dresselhaus, Creative Commons License.

which vanishes unless:

$$\vec{k} - \vec{k'} + \vec{G_m} = 0. \tag{5.48}$$

Since →' must lie inside the first Brillouin zone, the only difference that can exist betweenand →is the reciprocal lattice vector m 0. Consequently, Equation 5.47 becomes null unless →.

Furthermore, we have shown that the momentum matrix components connecting two bands can only exist at the same crystal momentum due to the periodicity of the lattice (Lee et al., 2016).

The momentum matrix elements that are utilized to calculate the effective mass in transport properties have a direct bearing on the probability of optical transitions. Consequently, learning about a solid's optical characteristics is essential to comprehending its transport characteristics (Navarrete et al., 2019).

5.6. INDIRECT INTERBAND TRANSITIONS

During indirect transitions, the semiconductor has the ability to either emit or absorb a phonon with an energy of :

$$\hbar\omega = E_f - E_i \pm \hbar\omega_q \tag{5.49}$$

The symbols and represent the energies of the initial and final electron states, respectively. The plus sign (+) indicates phonon emission, whereas the minus sign (-) indicates phonon absorption.

Now, let's derive an equation for the absorption coefficient in the case of indirect interband transitions in a semiconductor, considering the absorption of a phonon during the process. Here is a visual representation of this situation in Figure 5.9. Similar reasoning can be applied to situations involving the emission of a phonon (Yu & Fan, 2009).

The total process involves the utilization of the principle of energy conservation, which encompasses both the direct absorption of a phonon , and optical transition resulting in

$$\hbar\omega = E_g - \hbar\omega_q + \frac{\hbar^2(\vec{k}_n - \vec{k}_c)^2}{2m_n} + \frac{\hbar^2 k_p^2}{2m_p} \tag{5.50}$$

where: indicates the energy difference or thermal gap between the highest energy level in the valence band at the Γ point and the lowest energy level in the conduction band (for example, at the L point). The notation used in Equation 5.50 is defined in Figure 5.10. The phonon absorption process is indicated by the negative sign preceding the phonon energy in Equation 5.50. In Equation 5.50, represents the difference in crystal momentum hof an excited electron in the L-point conduction band compared to the minimum crystal momentum (Hassan & Spector, 1986). The kinetic energy of an excited electron with crystal momentum (h) equals:

$$E_n - E_c = \frac{\hbar^2(\vec{k}_n - \vec{k}_c)^2}{2m_n} \tag{5.51}$$

where: in Equation 5.51 is the effective mass of an electron close to the conduction band minimum; and is the energy above the conduction band minimum .

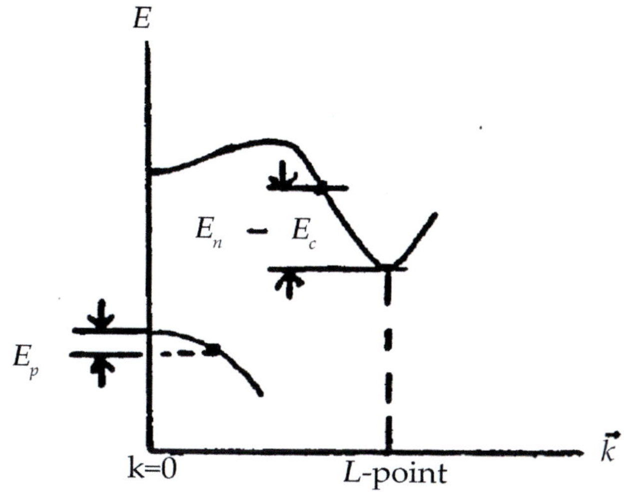

Figure 5.10. Schematic diagram of an indirect transition and the corresponding notation. is energy of conduction band at L point specifically at wave vector , while refers to the thermal energy gap.

Source: J.D. Jackson, Creative Commons License.

Given that the valence band extremum is located at = 0, the crystal momentum h represents the momentum of the hole generated upon electron excitation. The momentum is closely linked to the kinetic energy of the hole (Andersen & Bonderup, 2000).

$$E_p = \frac{\hbar^2 k_p^2}{2m_p}. \tag{5.52}$$

In this discussion, the sign convention is to consider as a positive value, and the reference point for energy is set at the valence band maximum (see to Figure 5.10). Regarding these sign conventions, the principle of conservation of energy results in:

$$\hbar\omega = E_g - \hbar\omega_q + (E_n - E_c) + E_p \tag{5.53}$$

(5.53) and conservation of momentum requires:

$$\vec{q} = \vec{k}_n - \vec{k}_p \tag{5.54}$$

where: the absorbed phonon's wave vector is denoted by .

The aim of our study is to investigate the relationship between frequency and the absorption edge for indirect transitions. By analyzing the optical absorption data frequency dependence, we will be able to distinguish between direct and indirect transitions. Now, let's examine the change from a certain starting state to a particular ending state (Gu et al., 2007). The energy dependence of the density of states () for the conduction band of the final state is given by:

$$\rho_c(E_n) \propto (E_n - E_c)^{1/2}. \tag{5.55}$$

By utilizing the conservation of energy equation stated in Equation 5.53, the function() can be represented in relation to .

$$\rho_c(E_n) \propto (\hbar\omega - E_g - E_p + \hbar\omega_q)^{1/2}. \tag{5.56}$$

Thus, it is noted that there are multiple initial states from which transitions to a state En can occur. This is because there are two extremes for the energy Ep: the extreme where all kinetic energy is transported to the hole () and the other extreme where = 0 (Pawlikowski et al., 1979). Within these bounds, the spectrum of possible valence band energies is represented by the energy δ.

$$\delta = \hbar\omega - E_g + \hbar\omega_q. \tag{5.57}$$

The energy dependence of the valence band's density of initial states is given by:

$$\rho_v(E_p) \propto E_p^{1/2} \tag{5.58}$$

where: the energy zero is defined by the standard 0, causing to disappear at the top of the valence band. For phonon absorption, the effective density of states can therefore be computed by adding up all of the energy-conserving values (Rafferty & Brown, 1998).

$$\rho(\hbar\omega) \propto \int_0^{\delta} \rho_c(E_n)\rho_v(E_p)dE_p \propto \int_0^{\delta} \sqrt{\delta - E_p}\,\sqrt{E_p}\,dE_p. \tag{5.59}$$

Equation 5.59 can be solved using integration by parts by defining the integration limits in terms of and using the notations u = , and v = .

$$\int_0^{\delta} \sqrt{uv}\,du = \frac{\delta - 2v}{4}\sqrt{uv}\,\bigg|_0^{\delta} + \frac{\delta^2}{4}\tan^{-1}\sqrt{\frac{u}{\delta - u}}\,\bigg|_0^{\delta} = \frac{\delta^2\pi}{8}. \tag{5.60}$$

When ⊠ in Eqs. 5.59 and 5.60 is substituted in Equation 5.57, the outcome is:

$$\rho(\hbar\omega) \propto \frac{\pi}{8}(\hbar\omega - E_g + \hbar\omega_q)^2 \tag{5.61}$$

It provides the indirect interband transition frequency dependency involving phonon absorption. Furthermore, the Bose-Einstein factor is directly proportional to the probability of a phonon being absorbed (Perlin et al., 2012).

$$n(\hbar\omega_q) = \frac{1}{\exp(\hbar\omega_q/k_BT) - 1} \tag{5.62}$$

so that in indirect transitions where a phonon is absorbed, the absorption coefficient becomes:

$$\alpha_{\text{abs}}(\omega) = C_a \frac{(\hbar\omega - E_g + \hbar\omega_q)^2}{\exp(\hbar\omega_q/k_BT) - 1} \tag{5.63}$$

where: the phonon absorption process is represented by a constant.

To compute the indirect absorption coefficient for phonon emission, the emission process's effective density of states must be determined (Lange et al., 1996). The only difference between this derivation and the last one for phonon absorption is that the phonon energy with a negative sign is now part of the energy conservation requirement (Catalano et al., 1988). Moreover, the emitting a phonon probability is directly related to , as expressed by:

$$[n(\hbar\omega_q) + 1] = 1 + [e^{\hbar\omega_q/k_BT} - 1]^{-1} = \frac{1}{1 - e^{-\hbar\omega_q/k_BT}} \tag{5.64}$$

such that the absorption coefficient for phonon emission is:

$$\alpha_{\text{ems}}(\omega) = C_e \frac{(\hbar\omega - E_g - \hbar\omega_q)^2}{1 - \exp(-\hbar\omega_q/k_BT)} \tag{5.65}$$

The symbol represents a constant that is specific to the process of phonon emission.

In low temperature conditions, the emission of phonons becomes the dominant process due to the scarcity of available phonons for absorption. In addition, when considering the energy of photons, distinct thresholds are observed for the absorption and emission processes. When it comes to the absorption process, absorption begins at a specific point, which is illustrated in Figure 5.11. On the other hand, for the emission process, optical absorption initiates at a different point (Danishevskii et al., 1970).

When we examine the graph in Figure 5.11, we can observe that the relationship between then in the low photon energy range follows the pattern . Similarly, is found to be proportional to . A combination of the emission and absorption processes will be found through observation. Experimental data showing indirect interband transitions are shown in Figure 5.12.

As the photon energy varies, the curves in Figure 5.12 shift, which is explained by the temperature dependency of the silicon indirect gap. Comparing Figures 5.11 and 5.12, it becomes apparent how the lower energy absorption contribution can be easily distinguished, as it is linked to the phonon absorption process. It is also simple to distinguish the contribution of phonon emission at higher energies. By conducting experiments at various temperatures, a more precise value for can be obtained (Cai et al., 2004).

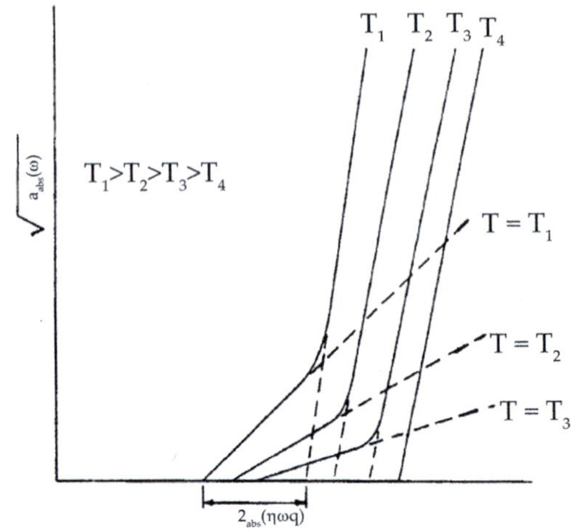

Figure 5.11. Schematic diagram showing the frequency dependency of coefficient of the absorption squared for indirect interband transitions around thresholds of absorption process and phonon emission. The curve indicates 4 different temperatures. The phonon emission mechanism is dominant at the lowest temperature T4, while at high temperature T1, the phonon absorption process is significant at low photon energies. It shows that photon energy is twice as large.

Source: G. Beke, Creative Commons License.

Figure 5.12 shows that phonon absorption improves with temperature, but emission is less influenced. In order for the absorption process to take place, it is necessary to have phonons with the appropriate wave vector. For , the process of phonon assistance involves the utilization of phonons with a wave vector that extends from Γ to L. On the other hand, in the case of Si, we require phonons with a vector that extends from Γ to (where the minimum energy level (Δmin) is linked with the lowest point in the conduction band at the Δ point). At lower temperatures, the number of phonons available is limited due to the thermal excitation of lattice vibrations. However, as the temperature increases, more phonons become accessible (Lancry et al., 2011).

In contrast, phonon emission is not influenced by the presence of phonons as it actually generates phonons during the emission process. As a result, the temperature has minimal impact on the phonon emission process (Hassan, 1994).

Due to its high hardness, silicon (with a Debye temperature of) only exhibits a small number of high-frequency phonons at ambient temperature. Hence, the mechanism of phonon emission will be the prevailing factor in optical absorption at photon energies that allow for such emission (Grzeszczyk et al., 2022). These theories can account for the differences in the slopes of the emission and phonon absorption contributions to the absorption coefficient in Figure 5.12.

One challenge with real materials is the existence of many phonon types for a given vector. There are specifically optical and acoustic branches, and longitudinal and transverse modes exist within each branch (Zatsepin et al., 2018).

Figure 5.13 provides an illustration of how optical absorption data can be analyzed to determine the frequencies of different phonons when q = 0. The plot shows the square root of vs hω for the indirect gap semiconductor GaP. This allows for the measurement of for distinct LO, LA, TO, and TA phonons. Inelastic neutron diffraction data is now used to correctly depict the dispersion curve for each phonon branch, making optical data collection rare (Erben et al., 2018).

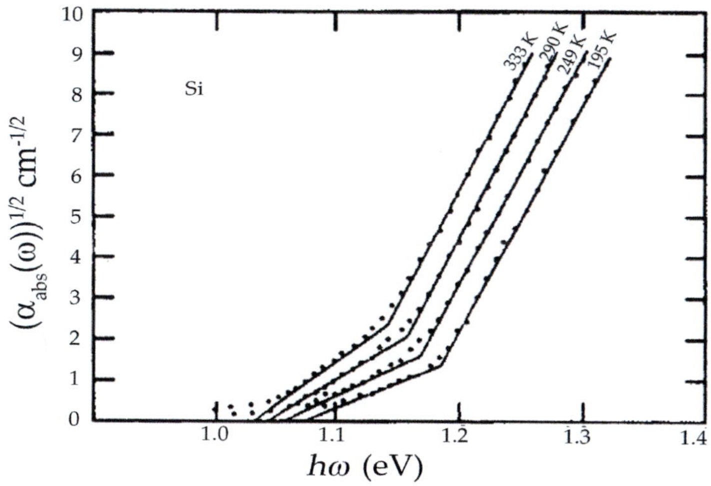

Figure 5.12. Plots showing the relationship between photon energy and the square root of the Si absorption coefficients at various temperatures. Two segments of a straight line drawn through experimental points show phonon emission and absorption contributions.

Source: J.D. Jackson, Creative Commons License.

ACTIVITY 5.1

Objective: To investigate the absorption characteristics of different semiconductor materials and understand how factors such as bandgap, temperature, and doping influence light absorption.

Materials Needed:

- Various semiconductor samples (e.g., silicon, gallium arsenide);
- Light source with adjustable wavelength;
- Spectrometer or photodetector;
- Temperature control apparatus (optional).

Procedure:

1. Measurement of Absorption Edge:

- Select a semiconductor sample and set up the experimental apparatus with a light source and spectrometer.
- Record the absorption spectrum by varying the wavelength of the incident light and noting the wavelengths where significant absorption occurs.
- Determine which absorption edge corresponds to the semiconductor's bandgap energy (E_g).

2. Temperature Dependence and Doping Effects:

- (Optional) Adjust the temperature of the semiconductor sample using a temperature control apparatus.
- Measure the absorption spectrum at different temperatures to observe changes in the absorption edge and discuss the temperature dependence of E_g.
- Alternatively, compare the absorption spectra of differently doped semiconductor samples to analyze how doping affects the absorption characteristics, particularly the absorption edge.

SUMMARY

- The chapter on Light Absorption in Solids offers an in-depth examination of the interaction between photons and solid-state materials, with a specific emphasis on fundamental mechanisms and phenomena. It begins with examining the Absorption Coefficient, a vital quantity that measures the efficiency with which a substance absorbs light at a specific wavelength.
- The chapter proceeds to explore Free Carrier Absorption, analyzing the process by which free electrons and holes in semiconductors and metals absorb photons through transitions between different energy levels. This phenomenon is crucial in comprehending the optical characteristics of these materials.
- The chapter delves deeper into the concept of Direct and Indirect Interband Transitions. These transitions occur when electrons absorb light energy and move between energy bands, either with the same crystal momentum states or different ones.
- It discusses the Temperature Dependence of the bandgap (E_g) and shows how temperature variations affect a material's optical characteristics. Furthermore, the Absorption Edge Dependence on Fermi Energy and Applied Electric Fields is investigated, emphasizing how changes in these factors affect materials' absorption properties.

REVIEW QUESTIONS

1. What is the absorption coefficient of a material and how is it determined?
2. Explain free carrier absorption in semiconductors and metals. How do their absorption mechanisms differ?
3. Define direct and indirect interband transitions in solids. How do they affect the optical properties of materials?
4. How does the band gap energy (Eg) of semiconductors vary with temperature? Provide a brief explanation.
5. Discuss the relationship between the absorption edge in semiconductors and the Fermi energy level. What implications does this relationship have?
6. Why is conservation of crystal momentum important in direct optical transitions?

MULTIPLE CHOICE QUESTIONS

1. The absorption coefficient of a material determines:

a. Its color under normal lighting conditions.

b. The efficiency of light absorption by the material.

c. The speed at which light travels through it.

d. Its ability to emit light when excited.

2. **Free carrier absorption in semiconductors occurs due to:**
   a. Absorption by bound electrons in valence bands.
   b. Absorption by lattice vibrations.
   c. Absorption by free electrons and holes.
   d. Absorption by impurity atoms.
3. **In metals, absorption of light primarily involves:**
   a. Absorption by phonons.
   b. Absorption by interstitial atoms.
   c. Absorption by free electrons.
   d. Absorption by excitons.
4. **Direct interband transitions in semiconductors refer to transitions where:**
   a. Crystal momentum is conserved.
   b. Crystal momentum is not conserved.
   c. There is a change in phonon momentum.
   d. The band gap changes with temperature.
5. **The temperature dependence of the band gap energy in semiconductors typically results in:**
   a. Decrease in band gap with increasing temperature.
   b. Increase in band gap with increasing temperature.
   c. Band gap remains constant regardless of temperature.
   d. Absence of band gap in semiconductors.
6. **The dependence of the absorption edge in semiconductors on the Fermi energy level primarily affects:**
   a. The electrical conductivity of the material.
   b. The color of light absorbed.
   c. The speed of light through the material.
   d. The mechanical strength of the material.

Answers to Multiple Questions

1. (b); 2. (c); 3. (c); 4. (a); 5. (b); 6 (a)

REFERENCES

1. Andersen, J. U., & Bonderup, E. (2000). Local field corrections for light absorption by fullerenes. The European Physical Journal D-Atomic, Molecular, Optical and Plasma Physics, 11, 435–448.

2. Aspnes, D. E. (1966). Electric-field effects on optical absorption near thresholds in solids. Physical Review, 147(2), 554.
3. Behzad, S. (2019). Direct to indirect band gap transition in two-dimensional germanium carbide through Si substitution. Results in Physics, 13, 102306.
4. Bonod, N., Tayeb, G., Maystre, D., Enoch, S., & Popov, E. (2008). Total absorption of light by lamellar metallic gratings. Optics express, 16(20), 15431–15438.
5. Cai, W. R., Mei, J. J., & Shen, W. Z. (2004). A proposal for complete interband absorption in indirect semiconductors. Physica B: Condensed Matter, 352(1–4), 179–184.
6. Catalano, I. M., Cingolani, A., Cingolani, R., & Lepore, M. (1988). Interband two-photon absorption mechanisms in direct and indirect gap materials. Physica scripta, 37(4), 579.
7. Danishevskii, A. M., Kastalskii, A. A., Ryvkin, S. M., & Yaroshetskii, I. D. (1970). Dragging of free carriers by photons in direct interband transitions in semiconductors. Sov. Phys. JETP, 31(2), 292–295.
8. Denninger, G., Dose, V., & Bonzel, H. P. (1982). Evidence for direct optical interband transitions in isochromat spectra from Pt single-crystal surfaces. Physical Review Letters, 48(4), 279.
9. Dou, X., Ding, K., Jiang, D., & Sun, B. (2014). Tuning and identification of interband transitions in monolayer and bilayer molybdenum disulfide using hydrostatic pressure. ACS nano, 8(7), 7458–7464.
10. Erben, D., Steinhoff, A., Gies, C., Schönhoff, G., Wehling, T. O., & Jahnke, F. (2018). Excitation-induced transition to indirect band gaps in atomically thin transition-metal dichalcogenide semiconductors. Physical Review B, 98(3), 035434.
11. Garriga, M., Cardona, M., Christensen, N. E., Lautenschlager, P., Isu, T., & Ploog, K. (1987). Interband transitions of thin-layer GaAs/AlAs superlattices. Physical Review B, 36(6), 3254.
12. Gobeli, G. W., Allen, F. G., & Kane, E. O. (1964). Polarization evidence for momentum conservation in photoelectric emission from germanium and silicon. Physical Review Letters, 12(4), 94.
13. Grzeszczyk, M., Novoselov, K. S., & Koperski, M. (2022). ZnPSe3 as ultrabright indirect band-gap system with microsecond excitonic lifetimes. Proceedings of the National Academy of Sciences, 119(41), e2207074119.
14. Gu, L., Srot, V., Sigle, W., Koch, C., van Aken, P., Scholz, F., ... & Rühle, M. (2007). Band-gap measurements of direct and indirect semiconductors using monochromated electrons. Physical Review B—Condensed Matter and Materials Physics, 75(19), 195214.
15. Guliamov, R., Lifshitz, E., Cohen, E., Ron, A., & Pfeiffer, L. N. (2001). Indirect barrier electron-hole gas transitions in mixed type-I–type-II GaAs/AlAs multiple quantum wells. Physical Review B, 64(3), 035314.
16. Hassan, A. R. (1994). Two-Photon Absorption in an Indirect-Gap Semiconductor Quantum Well System. I. Interband Transitions. physica status solidi (b), 184(2), 519–528.

17. Hassan, H. H., & Spector, H. N. (1986). Optical absorption in semiconducting quantum-well structures: Indirect interband transitions. Physical Review B, 33(8), 5456.
18. Hordvik, A., & Schlossberg, H. (1977). Photoacoustic technique for determining optical absorption coefficients in solids. Applied optics, 16(1), 101–107.
19. Jain, S. C., Nathan, A., Briglio, D. R., Roulston, D. J., Selvakumar, C. R., & Yang, T. (1991). Band-to-band and free-carrier absorption coefficients in heavily doped silicon at 4 K and at room temperature. Journal of applied physics, 69(6), 3687–3690.
20. James, R., & Smith, D. (1982). Theory of nonlinear optical absorption associated with free carriers in semiconductors. IEEE Journal of Quantum Electronics, 18(11), 1841–1864.
21. Jasieniak, J., Califano, M., & Watkins, S. E. (2011). Size-dependent valence and conduction band-edge energies of semiconductor nanocrystals. ACS nano, 5(7), 5888–5902.
22. Katahara, J. K., & Hillhouse, H. W. (2014). Quasi-Fermi level splitting and sub-bandgap absorptivity from semiconductor photoluminescence. Journal of Applied Physics, 116(17) , 2234–4454.
23. Koffyberg, F. P., Dwight, K., & Wold, A. (1979). Interband transitions of semiconducting oxides determined from photoelectrolysis spectra. Solid State Communications, 30(7), 433–437.
24. Lahbabi, M., Ahaitouf, A., Abarkan, E., Fliyou, M., Hoffmann, A., Charles, J. P., ... & Kerns, D. V. (2000). Analyses of electroluminescence spectra of silicon junctions in avalanche breakdown using an indirect interband recombination model. Applied Physics Letters, 77(20), 3182–3184.
25. Lambrecht, W. R. (2000). Electronic structure and optical spectra of the semimetal ScAs and of the indirect-band-gap semiconductors ScN and GdN. Physical Review B, 62(20), 13538.
26. Lancry, O., Farvacque, J. L., Pichonat, E., & Gaquière, C. (2011). Indirect interband transition in hexagonal GaN. Journal of Physics D: Applied Physics, 44(7), 075105.
27. Lange, H., Henrion, W., Fenske, F., Zettler, T., Schumann, J., & Teichert, S. (1996). Optical interband properties of some semiconducting silicides. physica status solidi (b), 194(1), 231–240.
28. Lee, B. G., Luo, J. W., Neale, N. R., Beard, M. C., Hiller, D., Zacharias, M., ... & Zunger, A. (2016). Quasi-direct optical transitions in silicon nanocrystals with intensity exceeding the bulk. Nano letters, 16(3), 1583–1589.
29. Lin, C. I., Baker, M., & Charlson, R. J. (1973). Absorption coefficient of atmospheric aerosol: A method for measurement. Applied Optics, 12(6), 1356–1363.
30. Linnros, J. (1998). Carrier lifetime measurements using free carrier absorption transients. I. Principle and injection dependence. Journal of Applied Physics, 84(1), 275–283.
31. Lyu, P., Espinoza, R., & Nguyen, S. C. (2023). Photocatalysis of Metallic Nanoparticles: Interband vs Intraband Induced Mechanisms. The Journal of Physical Chemistry C, 127(32), 15685–15698.

32. Malyi, O. I., & Acosta, C. M. (2020). Amorphization of indirect band gap semiconductors to tune their optoelectronic properties. The Journal of Physical Chemistry C, 124(27), 14432–14438.

33. Merino-Díez, N., Garcia-Lekue, A., Carbonell-Sanromà, E., Li, J., Corso, M., Colazzo, L., ... & de Oteyza, D. G. (2017). Width-dependent band gap in armchair graphene nanoribbons reveals Fermi level pinning on Au (111). ACS nano, 11(11), 11661–11668.

34. Miller, D. A., Chemla, D. S., Damen, T. C., Gossard, A. C., Wiegmann, W., Wood, T. H., & Burrus, C. A. (1985). Electric field dependence of optical absorption near the band gap of quantum-well structures. Physical Review B, 32(2), 1043.

35. Miller, D., Weiner, J., & Chemla, D. (1986). Electric-field dependence of linear optical properties in quantum well structures: Waveguide electroabsorption and sum rules. IEEE Journal of Quantum Electronics, 22(9), 1816–1830.

36. Mitra, C., & Lambrecht, W. R. (2008). Calculated interband optical transition spectra of GdN. Physical Review B—Condensed Matter and Materials Physics, 78(19), 195203.

37. Mott, N. F. (1938). On the absorption of light by crystals. Proceedings of the Royal Society of London. Series A. Mathematical and Physical Sciences, 167(930), 384–391.

38. Navarrete, F., Ciappina, M. F., & Thumm, U. (2019). Crystal-momentum-resolved contributions to high-order harmonic generation in solids. Physical Review A, 100(3), 033405.

39. Pace, C. N., Vajdos, F., Fee, L., Grimsley, G., & Gray, T. (1995). How to measure and predict the molar absorption coefficient of a protein. Protein science, 4(11), 2411–2423.

40. Pawlikowski, J. M., Misiewicz, J., & Mirowska, N. (1979). Direct and indirect optical transitions in Zn3P2. Journal of Physics and Chemistry of Solids, 40(12), 1027–1033.

41. Perlin, E. Y., Ivanov, A. V., & Popov, A. A. (2012). Interband phototransitions involving free electrons: I. Crystals with a direct band gap. Optics and Spectroscopy, 113, 376–382.

42. Rafferty, B., & Brown, L. M. (1998). Direct and indirect transitions in the region of the band gap using electron-energy-loss spectroscopy. Physical Review B, 58(16), 10326.

43. Rajkanan, K., Singh, R., & Shewchun, J. (1979). Absorption coefficient of silicon for solar cell calculations. Solid-State Electronics, 22(9), 793–795.

44. Šcepanovic, M. J., Grujic-Brojcin, M., Dohcevic-Mitrovic, Z. D., & Popovic, Z. V. (2007). Temperature dependence of the lowest frequency Eg Raman mode in laser-synthesized anatase TiO2 nanopowder. Applied Physics A, 86(3), 365–371.

45. Schroder, D. K., Thomas, R. N., & Swartz, J. C. (1978). Free carrier absorption in silicon. IEEE Journal of solid-state circuits, 13(1), 180–187.

46. Schumann Jr, P. A., & Phillips, R. P. (1967). Comparison of classical approximations to free carrier absorption in semiconductors. Solid-State Electronics, 10(9), 943–948.

47. Schwendel, D., Dahint, R., Herrwerth, S., Schloerholz, M., Eck, W., & Grunze, M. (2001). Temperature dependence of the protein resistance of poly-and oligo (ethylene glycol)-terminated alkanethiolate monolayers. Langmuir, 17(19), 5717–5720.

48. Segura, A., Cuscó, R., Attaccalite, C., Taniguchi, T., Watanabe, K., & Artús, L. (2021). Tuning the direct and indirect excitonic transitions of h-BN by hydrostatic pressure. The Journal of Physical Chemistry C, 125(23), 12880–12885.

49. Thomasson, S. I. (1980). On the absorption coefficient. Acta Acustica united with Acustica, 44(4), 265–273.

50. Vasquez, D. A. (1998). Optical transitions and momentum conservation in a one-dimensional crystal. European journal of physics, 19(4), 389.

51. Wang, L., Lai, S., & Liu, Y. (2022). Multi-frequency complete optical isolation based on indirect interband transition. Optical Engineering, 61(9), 097103–097103.

52. Wang, Y., Cong, C., Yang, W., Shang, J., Peimyoo, N., Chen, Y., ... & Yu, T. (2015). Strain-induced direct–indirect bandgap transition and phonon modulation in monolayer WS 2. Nano Research, 8, 2562–2572.

53. Yamaguchi, M., & Nobusada, K. (2016). Indirect interband transition induced by optical near fields with large wave numbers. Physical Review B, 93(19), 195111.

54. Yu, Z., & Fan, S. (2009). Complete optical isolation created by indirect interband photonic transitions. Nature photonics, 3(2), 91–94.

55. Zatsepin, A. F., Kuznetsova, Y. A., Mashkovtsev, M. A., & Rychkov, V. N. (2018). Interband optical transitions in Gd2O3: Er nanoparticles–prospective system for energy convertors. In IOP Conference Series: Materials Science and Engineering . IOP Publishing. 292(1) , 012047.

56. Zhang, X., Shi, G., Leveillee, J. A., Giustino, F., & Kioupakis, E. (2022). Ab initio theory of free-carrier absorption in semiconductors. Physical Review B, 106(20), 205203.

CHAPTER

6

Magnetic Properties of Atoms: Para and Dia-Magnetism

LEARNING OBJECTIVES

After studying this chapter, you will be able to:

- Understand paramagnet, diamagnet, and ferromagnet based on susceptibility and magnetization;
- Understand Hund's rules in atomic physics for electron orbital and spin configurations;
- Comprehend how Coulomb interactions drive spin alignment in atoms and molecules;
- Understand how electrons in atoms couple to external magnetic fields;
- Know the origins and effects of paramagnetic and diamagnetic terms in atomic Hamiltonians;
- Explain Larmor diamagnetism and its dependence on atomic properties;
- Describe the interplay between Pauli paramagnetism and Larmor diamagnetism in solids.

KEY TERMS FROM THIS CHAPTER

Curie paramagnetism	Diamagnetism
Ferromagnetism	Hund's rules
Landau diamagnetism	Larmor diamagnetism
Magnetization	Paramagnetism
Pauli paramagnetism	Susceptibility

UNIT INTRODUCTION

One might wonder why there is an interest in magnets. Although magnetism has been observed since ancient times, our understanding of its underlying causes has significantly advanced with the advent of quantum mechanics. This particular area of physics, which one will be exploring in-depth, offers a unique opportunity to witness the fascinating interplay between statistical physics and quantum physics. The band theory regarding electrons falls short when attempting to explain the behavior of magnets. Indeed, the study of magnetism continues to be a highly dynamic field in physics, with numerous challenging and unresolved inquiries (Wilsdon, 1925). Condensed matter physics often relies on magnetism to explore and gain insights into the intricate world of quantum and statistical physics, both in theoretical studies and experimental settings.

It is important to note that the majority of magnetic phenomena are a result of the quantum-mechanical characteristics exhibited by electrons. Although nuclei possess magnetic moments and may add to magnetic fields, the magnitude of these nuclear moments is generally smaller compared to that of electrons (Nesterenko et al., 2003).

Discover the underlying principles behind the magnetic properties exhibited by atoms in this unit. The discussion starts by explaining different forms of magnetism and the way the magnetic poles of electrons in atoms align. Explore the fascinating world of paramagnetism along with diamagnetism, where materials showcase their magnetic behavior when influenced by external fields of magnets. The unit also explores Hund's rules within atomic physics, which dictate electron layouts, and investigates specific phenomena observed in metals, such as Larmor diamagnetism and Pauli paramagnetism (Van Vleck, 1928).

6.1. BASIC DEFINITIONS OF TYPES OF MAGNETISM

Let us first make some definitions. Recall that for a small magnetic field, the magnetization of a system M (moment per unit volume) is typically related linearly to the applied magnetic field H by a (magnetic) susceptibility χ. (Dekkers, 1997). We write for small fields H,

$$M = \chi H \quad (6.1)$$

It is important to note that χ seems a dimensionless quantity. In cases where susceptibilities are small (which is usually the case, except in ferromagnets), there is minimal distinction between H and B (where represents the permeability for free space). Therefore, one can also express this as:

$$M = \chi B / \mu_0. \quad (6.2)$$

- A paramagnet seems a material that $\chi > 0$ exhibits positive susceptibility, meaning that its resulting magnetization aligns with the applied magnetic field (Néel, 1955);
- A diamagnet represents a material that $\chi < 0$ exhibits a negative susceptibility, meaning that its magnetization opposes the direction of the applied magnetic field;
- A ferromagnet seems a material in which M can have a non-zero value, even without the presence of any external magnetic field (Brown et al., 2002).

Diamagnetism is a common occurrence that typically takes place unless it is overshadowed by other magnetic phenomena. Take water, along with most other biological substances, for instance. That exhibit diamagnetic properties. From a qualitative perspective, one can draw a parallel between diamagnetism and Lenz's law. According to this principle, an induced current creates a magnetic field that opposes the alteration that caused it (Inoue et al., 1981). Nevertheless, the comparison is not exact. When a magnetic field is present, a loop of wire can produce a current that produces magnetization in the opposite direction. Nevertheless, in a loop of wire that is not superconducting, the electrical current will gradually diminish until it reaches zero, resulting in the absence of any left magnetization. Within a diamagnet, the magnetization persists as long as the magnetic field that is applied has been present (Mulders et al., 1998).

6.2. ATOMIC PHYSICS: HUND'S RULES

One begins by examining the basic principles of electrons within a solitary atom, disregarding the fact that atoms within materials are not solitary but rather connected to other atoms. In basic quantum mechanics, it is known that an electron entering an atomic orbital may be identified using four quantum numbers: n. The majority of elements possess many complete shells and one incomplete shell, which is referred to as the valence shell. To comprehend magnetism in circumstances when shells are only partially filled, it is necessary to determine which of the accessible () orbitals are in use within these shells, as well as the filled spin states (). Specifically, if these electrons possess a collective magnetic orientation (Pyper & Grant, 1977). There are certain rules, referred to as "Hund's Rules," that dictate how electrons complete orbitals and figure out their spin states. These rules ultimately determine whether an isolated atom possesses a magnetic moment. One way to demonstrate these rules is by using a specific example. As stated earlier, this element in its atomic form includes three electrons within its outermost shell. This shell is known as an f-shell, indicating that it possesses angular momentum with a value of l = 3. Consequently, it has a total of 7 potential values for (). Additionally, each electron can have two possible spin values (Colpa & Brown, 1973). So, where should we place the three electrons among these potential orbital/spin states?

- "Hund's First Rule (Paraphrased): Electrons try to align their spins." Based on this rule, it is understood that all three valence electrons within Pr will align their spins in the same direction. As a result, one obtains an overall spin angular momentum S = 3/2 using the three S = 1/2 spins. On a local scale, specifically within a single atom, the spins of the three electrons exhibit a ferromagnetic behavior as they all align (Liu & Langenaeker, 2003). Now, one needs to determine the orbital states for placing the electrons. Another rule is required for this:
- "Hund's Second Rule (Paraphrased): Electrons try to maximize their total orbital angular momentum, consistent with Hund's First Rule." : When considering Pr, one completes the = 3, = 2, and = 1 states to achieve the highest possible total value of 6. This results in an overall L value of 6, and due to rotational invariance, one can effectively guide L in all directions. Therefore, one complete orbital is illustrated in Figure 6.1. In the diagram, one has positioned the spins towards the right to optimize , following Hund's second rule. Additionally, one aligned all the spins, following Hund's first rule. It is important to consider

that placing two electrons within the same orbital is not possible, as they must have spin alignment and adhere to the Pauli principle. Once more, the principle of optimizing orbital angular momentum can be affected by the physics for Coulomb interaction (Morgan & Kutzelnigg, 1993).

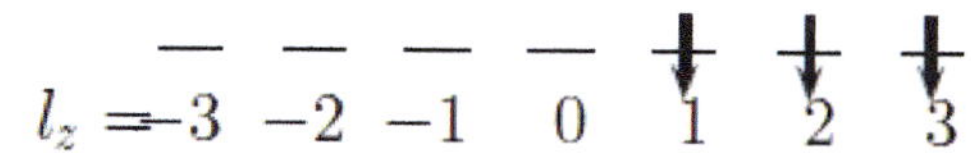

Figure 6.1. The filling in the f shell for a Pr atom is by Hund's rules. We align the spins and maximize the component to maximize the total angular momentum, L.

Source: Steven H, Creative Commons License.

Currently, one has S = 3/2 or L = 6. However, it is important to consider the alignment of spin or orbital angular momenta to each other. Now, let's move on to the last rule:

"Hund's Third Rule (paraphrased): Given Hund's first and second rules, the orbital and spin angular momentum either align or antialign, so that the total angular momentum is J = L S with the sign being determined by whether the shell of orbitals is more than half filled (+) or less than half filled (–)." The explanation for this rule cannot be due to interaction physics but rather spin–orbit coupling. The Hamiltonian usually contains spin-orbit terms $\alpha\ l\ \sigma$, along with the sign of α defines the alignment of spin and orbit to reduce energy. Therefore, in the case of Pr, with an electron configuration of J = L S = 9/2, and a partially filled shell, the overall angular momentum of J is equal to 9/2. It is important to note that J is often colloquially referred to as the "spin" of the atom. This is a commonly used term that is often used but not very precise (Hongo et al., 2004). It is important to note that J represents the overall angular momentum for the electrons within the atom, while S specifically refers to the spin aspect of J (Natoli & McMillin, 2018).

6.2.1. Moments Alignment

Now, one will delve into the reasons behind the effectiveness of Hund's rules. Specifically, one will be interested in understanding why magnetic moments, whether they are real spin moments or orbital moments, tend to align with each other. This section will provide a qualitative analysis, offering a general understanding of the underlying physics (Zellinger et al., 2019).

Now, turning attention to Hund's first rule and exploring why spins tend to align. To begin with, it is important to note that there is no connection to magnetic dipole interactions. Although the interactions between the magnetic dipoles and their spins occur, the energy scale becomes extremely small and insignificant when the dipole moments fall within the order from the Bohr magneton. Instead, the alignment arises due to Coulomb interaction energy. Now explore the mechanics by examining the wave function of two electrons within an atom (Dong & Boyer, 1996).

6.2.1.1. Naive Argument

Pauli's exclusion principle requires the whole wave function to be antisymmetric. It is typically possible to express:

$$\Psi(r_1,\sigma_1;r_2,\sigma_2)=\psi_{orbital}(r_1,r_2)\chi_{spin}(\sigma_1,\sigma_2)$$

The variables represent the locations of the particles, while represents their spin. If both spins align, namely both being spin-up (i.e., () = 1 and = 0 for all other spin setups), their spin wave function becomes symmetric, whereas the spatial waves function has to be antisymmetric (Hample et al., 1999). Consequently, one has:

$$\lim_{r_1\to r_2}\psi_{orbital}(r_1,r_2)\to 0.$$

When spins of electrons are aligned, these particles are unable to come into proximity to each other, resulting in a decrease in the system's Coulomb energy (Aakhus, 2003).

6.2.1.2. Exchange Energy

The energy that separates having two spins lined up versus anti-aligned is commonly referred to as the two-way interaction or as exchange energy. Those familiar with atomic physics will remember that physicists use the term "exchange" to describe the hopping matrix element, which involves the transfer of an electron between orbitals (Becke, 1988). Interestingly, the present name is strongly connected. Exploring the relationship by performing a straightforward calculation of the energy variance among two electrons with aligned spins or two electrons with anti-aligned spins. Now consider two electrons located on separate orbitals, which one will refer to as A(r) and B(r). The general wave function is written I, or to be antisymmetric, which can be expressed as the absolute difference between AB and BA. Alternatively, if one chooses for spins that are antialigned (a singlet, hence antisymmetric, i.e., the spatial wavefunction has to be symmetric. By incorporating the Coulomb interaction V (r1, r2) (Sahni et al., 1982), one can determine the energy disparity between the singlet and triplet states through the following calculation:

$$\begin{aligned}
E_{singlet} &= (\langle AB| - \langle BA|)V(|AB\rangle - |BA\rangle) \\
E_{triplet} &= (\langle AB| + \langle BA|)V(|AB\rangle + |BA\rangle) \\
E_{exchange} &= E_{singlet} - E_{triplet} = -4\,\mathrm{Re}\langle AB|V|BA\rangle.
\end{aligned}$$

Figure 6.2 illustrates the wave function for a single electron, while the other electron (located to the left) is shown as being in a stationary state.

- Top: In the context of electron spin, the presence of another electron with opposite spin results in a reduction of the effective charge experienced by the original electron within the nucleus. This phenomenon can be attributed to the screening effect (Barrow & Clifton, 2006);
- Bottom: On the other hand, once spins align themselves, the two electrons are unable to approach each other, causing the fixed electron to perceive a greater charge from the nucleus. The binding between the fixed electron and the nucleus is enhanced if the two electrons have aligned spins, leading to a configuration with lower energy (Murray et al., 2009).

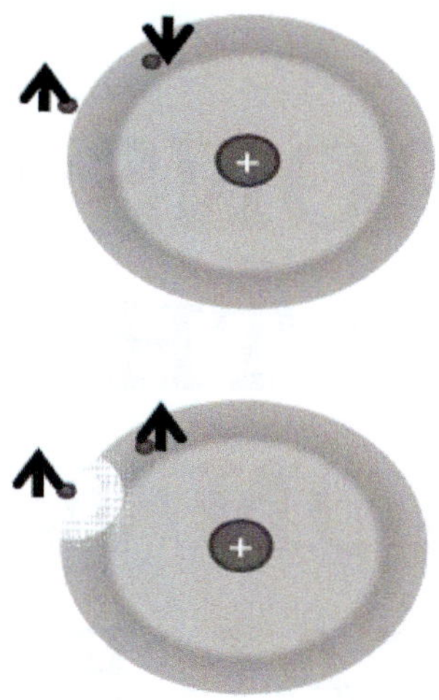

Figure 6.2. Illustration of Hund's first rule.

Source: Simon L, Creative Commons License.

6.2.1.3. Magnetic Interactions in Molecules and Solids

It is important to exercise caution when dealing with such kinds of arguments, especially when it is applied to molecules rather than individual atoms. When considering a diatomic molecule like , there are two electrons along with two nuclei involved. Although the effect of screening (Figure 6.2) still takes place and attempts to align both electrons, its effectiveness is slightly reduced compared to two electrons upon a single atom, as both of the electrons are often found near opposite nuclei almost all the time (Szilva et al., 2023). In addition, there is a conflicting influence that causes the electrons have to align in opposite directions. In this large two-atom box, there exists a lowest-energy (symmetric) wave function. The most favorable state for two electrons in terms of their lowest energy is going to have their spins anti-aligned. This allows both electrons to occupy an identical low-energy spatial wave function. It can be quite challenging to determine the alignment preferences of electrons on neighboring atoms (Novoa et al., 2011).

6.3. ELECTRONS IN ATOMS AND THEIR INTERACTION WITH AN EXTERNAL FIELD

After exploring the alignment of electron moments, now delve into the interaction between electrons in atoms and an external magnetic field. When there is no magnetic field present, the Hamiltonian over an electron in an atom follows the typical pattern, with V representing the electrostatic potential generated by the nucleus and potentially other electrons. Now explore the effects of introducing an outside magnetic field (Dzuba et al., 2018).

$$H_0 = \frac{p^2}{2m} + V(r)$$

Remember that the Hamiltonian overcharged particles within a magnetic field B is expressed in the limited coupling form:

$$H = \frac{(p+eA)^2}{2m} + g\mu_B B \cdot \sigma + V(r)$$

In the given equation, -e represents the charge attached to the particle (specifically, the electron), σ denotes the electron spin, g represents the electron g-factor (which is approximately 2), is the Bohr magneton defined as , and A represents the vector potential. In the case of a magnetic field (uniform), one can express, resulting in. Next, one has:

$$H = \frac{p^2}{2m} + V(r) + \frac{e}{2m} p \cdot (B \times r) + \frac{e^2}{2m}\frac{1}{4} | B \times r |^2 + g\mu_B \cdot \sigma$$

The initial two terms within this equation represent the Hamiltonian when there is no external magnetic field (Smirnov & Chibisov, 1966). Now rephrase the following term:

$$\frac{e}{2m}\mathrm{p}\cdot(\mathrm{B}\times\mathrm{r})=\frac{e}{2m}\mathrm{B}\cdot(\mathrm{r}\times\mathrm{p})=\mu_B\mathrm{B}\cdot\mathrm{l} \quad (6.4)$$

The equation = r × p represents the angular momentum on the orbit given by the electron. This can be combined with the Zeeman terms to obtain:

$$\mathcal{H}=\mathcal{H}_0+\mu_B\mathrm{B}\cdot(\mathrm{l}+g\sigma)+\frac{e^2}{2m}\frac{1}{4}|\mathrm{B}\times\mathrm{r}|^2 \quad (6.5)$$

The part located in the middle on the right side of this equation, referred to by the paramagnetic term, represents the interaction between the outer field and the combined magnetic field of the electron, which includes both the orbital moment and the spin moment It's important to observe that when a magnetic field is used, the moments of these particles align themselves with the field (while l and σ align in the opposite direction) to decrease the energy. Due to the influence of an external field, a moment is generated in the same direction, leading to the occurrence of paramagnetism (Krems, 2005).

The last term of Equation 6.5 is referred to be the diamagnetic terms in the Hamiltonian, and it is responsible for causing the phenomenon of diamagnetism. The quadratic term within B will consistently lead to a rise in the atom's total energy whenever a magnetic field is present. As a result, it has a reverse impact compared to the paramagnetic term discussed earlier. Given the quadratic nature of this term in B, it is reasonable to anticipate that it would have less significance compared to a paramagnetic phrase (that is straight in B) when B is small (Lambropoulos et al., 1998).

The Hamiltonian contains two terms that contribute to the response of atoms toward external magnetic fields, resulting in both paramagnetic and diamagnetic behavior. One will discuss each of them separately in the following two sections. It's important to note that the current focus is on the magnetic behavior of an individual atom (Schwinger, 1948).

6.4. FREE SPIN (CURIE OR LANGEVIN) PARAMAGNETISM

Let's begin by examining the impact of the paramagnetic terms in Equation 6.5. Assuming the problem of the unaffected Hamiltonian is now solved, one can focus solely on the redirection of the spin σ along with the orbital angular momentum for an electron, without needing to consider the rest of the Hamiltonian. At this point, we also disregard the diamagnetic term of the Hamiltonian, as its effect is generally weaker than that of the paramagnetic term (Nguyen et al., 1996).

6.4.1. Free Spin 1/2

Now take a moment to revisit a simpler scenario that one may recall from studies in mathematical physics: a free spin-1/2. As one might remember, the Hamiltonian for a single spin-1/2 can be expressed as:

$$H = g\mu_B B \cdot \sigma \quad (6.6)$$

Given that the g-factor for the spin is set as 2, and μB = eℏ/(2m) represents the Bohr magneton. This can be considered as a simplified form for the paramagnetic term described in Equation 6.5, specifically for a solitary unbound electron, disregarding the orbital angular momentum. The eigenstates of B σ were ±B/2, resulting in a partition function (Ramirez et al., 1994).

$$Z = e^{-\beta\mu_B B} + e^{\beta\mu_B B} \quad (6.7)$$

Furthermore, one can derive the moment of magnetism (per spin) by utilizing the associated free energy log Z.

$$moment = -\frac{\partial F}{\partial B} = \mu_B \tanh\left(\beta\mu_B B\right). \quad (6.8)$$

When multiple atoms are present in a given space, the magnetization M can be defined as the magnetic moment divided by the volume. Next, when considering a small field one can calculate the susceptibility as:

$$\chi = \lim_{H \to 0} \frac{\partial M}{\partial H} = \frac{n\mu_0 \mu_B^2}{k_B T} \quad (6.9)$$

In this case, the value of represents the total number of spins for each volume. It is worth noting that one has made use of, where is the permeability that is of free space. Expression 6.9 is commonly referred to as the "Curie law" susceptibility. Any susceptibility from the form over a constant C is also known as Curie law. Paramagnetism including free spins, such as this, can be referred to as Curie paramagnetism or Langevin paramagnetism (Griffiths, 1964).

6.4.2. Free Spin J

The paramagnetic term within the Hamiltonian can be more complex than the spin-1/2 model described in Equation 6.6. On the other hand, when one analyzes Equation 6.5 and extend it to multiple electrons within an atom, it becomes apparent that one must take into account a Hamiltonian with a specific structure (Freeman & Hill, 1971).

$$\mathcal{H} = \mu_B \mathbf{B} \cdot (\mathbf{L} + g\mathbf{S}) \quad (6.10)$$

The combined orbital or spin components that make up the electrons within the atom are denoted by L and S. Remember that Hund's rules provide us with the values of L, S, and J. The equation in Equation 6.10 appears to be somewhat inconvenient, as Hund's third rule provides information about J = L+S rather than L + gS. Luckily, the matrix elements one is interested in (reorientations of J lacking altering the values of J, S, or L as determined by Hund's rules) (Lloyd & Pake, 1954) are exactly equivalent to the Hamiltonian Equation 6.10.

$$\mathcal{H} = \tilde{g}\mu_B \mathbf{B} \cdot \mathbf{J} \quad (6.11)$$

where: represents the effective g-factor, also known to be the Landé g-factor, which is defined as:

$$\tilde{g} = \frac{1}{2}(g+1) + \frac{1}{2}(g-1)\left[\frac{S(S+1) - L(L+1)}{J(J+1)}\right]$$

Constructing the partition function becomes easy with the new Hamiltonian.

$$Z = \sum_{J_z=-J}^{J} e\beta\tilde{g}\mu_B B J_z. \quad (6.12)$$

Just like the spin-1/2 scenario stated earlier, one can differentiate to find the moment as a function for temperatures. When examining the density of such atoms, one can observe the magnetization and susceptibility (Griffiths, 1964). As per the Curie form, the result provides the degree of susceptibility per unit volume, as stated in Equation 6.9.

$$\chi = \frac{n\mu_0(\tilde{g}\mu_B)^2}{3}\frac{J(J+1)}{k_BT}.$$

It is worth noting that the susceptibility of the Curie law shows a significant increase at low temperatures. If the term was non-zero (i.e., if J is non-zero), then the Curie paramagnetism prevails over other forms of paramagnetism or diamagnetism (Matsen, 1964).

6.5. LARMOR DIAMAGNETISM

When J = 0, Curie paramagnetism becomes the dominant factor, making it unlikely to observe diamagnetism in atoms. An interesting scenario arises when dealing with atoms that have filled shell configurations, such as noble gases, where L = S = J = 0. There is another possibility where J could be zero, even if L and S are non-zero. Regardless of the situation, the expectation of the paramagnetic term in Equation 6.5 is zero, so it can be largely disregarded. It is important to take into account the impact of the final equation in Equation 6.5, known as the diamagnetic term (Szmytkowski, 2002).

Suppose one considers the application of B within the z^ direction. The Hamiltonian's diamagnetic term (Equation 6.5) can then be expressed as the expectation value.

$$\delta E = \frac{e^2}{8m}\left\langle |B \times r|^2 \right\rangle = \frac{e^2 B^2}{8m}\left\langle x^2 + y^2 \right\rangle$$

By leveraging the atom's rotational symmetry, one can express:

$$\left\langle x^2 + y^2 \right\rangle = \frac{2}{3}\left\langle x^2 + y^2 + z^2 \right\rangle = \frac{2}{3}\left\langle r^2 \right\rangle$$

so, one has:

$$\delta E = \frac{e^2 B^2}{12m}\left\langle r^2 \right\rangle$$

Therefore, the magnetic moment for each electron is:

$$moment = -\frac{dE}{dB} = -\left[\frac{e^2}{6m}\left\langle r^2 \right\rangle\right] B$$

When considering the density of electrons in a system, it is possible to express the susceptibility in a certain way.

$$x = \frac{\rho e^2 \mu_0 \left\langle r^2 \right\rangle}{6m} \qquad (6.13)$$

The result, Equation 6.13, is commonly referred to as Larmor diamagnetism. Typically, the value of for atoms is approximately some Bohr radii squared. Interestingly, a similar concept can be extended to large conductive molecules, where the electrons have the freedom to move along the entire length of the molecule. In this case, the value would represent the squared radius of the molecule, rather than that of the atom (Essén, 1989).

Remember

Some materials, like bismuth and graphite, generate a magnetic field that goes against an external magnetic field, causing a mild repulsion. This distinct characteristic arises from the creation of induced currents within the material, which in turn produce a magnetic field that opposes.

6.6. ATOMS IN SOLIDS

So far, one has only been discussing the magnetism (paramagnetism or diamagnetism) of individual atoms in isolation. While the atomic model provides a solid understanding of magnetism, the dynamics in solids can present some variations (Moroni et al., 1997).

6.6.1. Pauli Paramagnetism in Metals

One determined the susceptibility for a free Fermi gas by using , which represents the density of states within the Fermi surface (Fujiyama et al., 2022).

$$x_{Pauli} = \mu_0 \mu_B^2 g(E_F) \qquad (6.14)$$

It is reasonable to anticipate that this formulation would apply to metals with complex band structures, despite the only alteration being the adjustment of the density of states. Undoubtedly, this statement is quite accurate when applied to basic metals like Li or Na (Maki, 1966).

It's interesting to observe that the susceptibility for each spin of a Fermi gas (Equation 6.14) is roughly one factor smaller in comparison to the susceptibility over a free spin (Equation 6.9). It is important to understand that, because of the Pauli exclusion principle, just a tiny fraction of spins close to the Fermi surface may be flipped, resulting in a minimal susceptibility (Bergmann, 1978).

6.6.2. Diamagnetism in Solids

One applied our calculation for Larmor diamagnetism to isolated atoms with J = L = S = 0, specifically noble gas atoms. At low temperatures, noble gas atoms can come together to create loosely bonded crystals. This behavior is consistent across most noble gases, except for helium, which instead transforms into a superfluid in low temperatures. When Equation 6.13 is applied to noble gas crystals, its density of electrons ⍰ is set to be equivalent to the density of atoms n multiplied by the total number of electrons for each atom Z (Bakhmutov, 2011). Therefore, when it comes to atoms of noble gases, one obtains:

$$x_{Larmor} = \frac{Zne^2 \mu_0 \left\langle r^2 \right\rangle}{6m} \qquad (6.15)$$

where: the value of is determined through the atomic radius.

For any substance, the Hamiltonian includes a diamagnetic term that accounts for the coupling of orbital motion with a magnetic field. As a result, there will always be a certain level of diamagnetism present. Equation 6.15 is typically quite accurate in explaining the diamagnetism for electrons within core orbitals. The calculation of the conduction electrons within a metal is more complex, resulting in what is known as Landau diamagnetism (Zhao et al., 2024). This phenomenon, when combined with Pauli paramagnetism, leads to a reduction in the overall paramagnetism of the conduction electrons by 1/3.

$$\chi_{Landau} = -\frac{1}{3}\chi_{Pauli}$$

When examining a metal such as copper, it may seem logical to assume that it would exhibit paramagnetic behavior, taking into account the Pauli paramagnetism mentioned earlier (which is influenced due to the Landau effect) (Ghosal et al., 2007). Interestingly, copper exhibits diamagnetic properties. There is a possible explanation for this phenomenon, which involves the core electrons in copper having strong Larmor diamagnetism that can counteract the Pauli paramagnetism exhibited by the conduction electrons! It is worth noting that in metals, Larmor diamagnetism can be quite powerful, sometimes even surpassing Pauli paramagnetism. This is especially evident in heavy elements, where the presence of numerous core electrons enhances the diamagnetic effect. It is worth noting that in materials with free spins, Curie paramagnetism remains stronger compared to any diamagnetism (Sondheimer & Wilson, 1951).

ACTIVITY 6.1

Objective: To observe and classify different materials based on their magnetic properties.

Materials Needed:

- Various magnets (different sizes and types);
- Samples of materials (metals, non-metals, insulators).

Activity Steps:

1. Demonstration
    - Show participants the magnets and different materials.
    - Demonstrate how each material interacts with magnets:
    - Test attraction or repulsion with magnets.
    - Classify materials as paramagnetic (attracted), diamagnetic (weakly repelled), or non-magnetic (no interaction).
2. Conclusion:
    - Summarize findings and discuss practical applications of understanding magnetic properties in everyday materials.

SUMMARY

- The chapter provides a comprehensive exploration of magnetism at the atomic level. It begins with foundational definitions of magnetism types and delves into atomic physics principles such as Hund's rules and moment's alignment.
- The discussion progresses to elucidate how electrons in atoms respond to external magnetic fields, covering phenomena like free spin (Curie or Langevin) paramagnetism for both spin ½ and higher spin systems.
- Additionally, the chapter addresses Larmor diamagnetism and examines the behavior of atoms within solids, particularly highlighting Pauli paramagnetism in metals and diamagnetism in solid materials.

REVIEW QUESTIONS

1. Explain the differences between paramagnetism and diamagnetism. Provide examples of materials exhibiting each type of magnetism.
2. How do Hund's rules influence the arrangement of electrons in atoms? Discuss their relevance to magnetic properties.
3. Describe the concept of free spin paramagnetism. What factors influence the magnitude of paramagnetic susceptibility in materials?
4. What is Larmor diamagnetism? How does it differ from other forms of magnetism such as paramagnetism and diamagnetism?
5. Discuss the role of Pauli paramagnetism in metals. How does it contribute to the overall magnetic behavior of metallic materials?
6. How do atoms in solids behave differently magnetically compared to isolated atoms? Provide examples to illustrate your answer.
7. Explain the phenomenon of magnetic moments alignment in the presence of an external magnetic field. How does this alignment affect the magnetic properties of materials?

MULTIPLE CHOICE QUESTIONS

1. Which of the following is a characteristic of paramagnetic materials?

a. They have a net magnetic moment even in the absence of an external magnetic field.

b. They are strongly repelled by magnetic fields.

c. Their magnetic moments align parallel to an applied magnetic field.

d. They exhibit zero magnetic susceptibility.

2. Hund's rules in atomic physics primarily describe:

a. Electron configuration within atoms.

b. Behavior of neutrons in magnetic fields.

c. Interaction between protons and electrons.

d. Magnetic alignment of nuclei.

3. **Which type of magnetism arises from the alignment of unpaired electron spins in the presence of an external magnetic field?**

   a. Diamagnetism

   b. Paramagnetism

   c. Ferromagnetism

   d. Antiferromagnetism

4. **Larmor diamagnetism is primarily observed in:**

   a. Solids with no free electrons.

   b. Non-magnetic materials.

   c. Metals at high temperatures.

   d. Paramagnetic materials.

5. **Pauli paramagnetism in metals results from:**

   a. Spontaneous alignment of electron spins.

   b. Coupling of electron spins.

   c. Repulsion between electron spins.

   d. Interaction with external magnetic fields.

Answers to Multiple Questions

1. (c); 2. (a); 3. (b); 4. (b); 5. (d)

REFERENCES

1. Aakhus, M. (2003). Neither naïve nor critical reconstruction: Dispute mediators, impasse, and the design of argumentation. Argumentation, 17, 265–290.
2. Bakhmutov, V. I. (2011). Strategies for solid-state NMR studies of materials: from diamagnetic to paramagnetic porous solids. Chemical Reviews, 111(2), 530–562.
3. Barrow, J. D., & Clifton, T. (2006). Cosmologies with energy exchange. Physical Review D—Particles, Fields, Gravitation, and Cosmology, 73(10), 103520.
4. Becke, A. D. (1988). Density-functional exchange-energy approximation with correct asymptotic behavior. Physical review A, 38(6), 3098.
5. Bergmann, G. (1978). Transition from Pauli paramagnetism to band ferromagnetism in very thin Ni films. Physical Review Letters, 41(4), 264.
6. Brown, D., Ma, B. M., & Chen, Z. (2002). Developments in the processing and properties of NdFeb-type permanent magnets. Journal of magnetism and magnetic materials, 248(3), 432–440.
7. Colpa, J. P., & Brown, R. E. (1973). Hund's rules and the interpretation and common misinterpretation of energy differences. Molecular Physics, 26(6), 1453–1463.

8. Dekkers, M. J. (1997). Environmental magnetism: an introduction. Geologie en mijnbouw, 76, 163–182.
9. Dong, L., & Boyer, A. L. (1996). A portal image alignment and patient setup verification procedure using moments and correlation techniques. Physics in Medicine & Biology, 41(4), 697.
10. Dzuba, V. A., Berengut, J. C., Ginges, J. S. M., & Flambaum, V. V. (2018). Screening of an oscillating external electric field in atoms. Physical Review A, 98(4), 043411.
11. Essén, H. (1989). Magnetic fields, rotating atoms and the origin of diamagnetism. Physica Scripta, 40(6), 761.
12. Freeman, R., & Hill, H. D. W. (1971). High-Resolution Study of NMR Spin Echoes:"J Spectra". The Journal of Chemical Physics, 54(1), 301–313.
13. Fujiyama, S., Maebashi, H., Tajima, N., Tsumuraya, T., Cui, H. B., Ogata, M., & Kato, R. (2022). Large diamagnetism and electromagnetic duality in two-dimensional Dirac electron system. Physical Review Letters, 128(2), 027201.
14. Ghosal, A., Goswami, P., & Chakravarty, S. (2007). Diamagnetism of nodal fermions. Physical Review B—Condensed Matter and Materials Physics, 75(11), 115123.
15. Griffiths, R. B. (1964). A proof that the free energy of a spin system is extensive. Journal of Mathematical Physics, 5(9), 1215–1222.
16. Hample, D., Benoit, P. J., Houston, J., Purifoy, G., VanHyfte, V., & Wardwell, C. (1999). Naive theories of argument: Avoiding interpersonal arguments or cutting them short. Argumentation and Advocacy, 35(3), 130–139.
17. Hongo, K., Maezono, R., Kawazoe, Y., Yasuhara, H., Towler, M. D., & Needs, R. J. (2004). Interpretation of Hund's multiplicity rule for the carbon atom. The Journal of chemical physics, 121(15), 7144–7147.
18. Inoue, M., Oshima, H., Morisaki, M., Yagi, H., Fun, H. K., & Tatsukawa, T. (1981). Various types of magnetism in SnTe crystals doped with 3d transition metals. Journal of the Physical Society of Japan, 50(4), 1222–1226.
19. Krems, R. V. (2005). Molecules near absolute zero and external field control of atomic and molecular dynamics. International Reviews in Physical Chemistry, 24(1), 99–118.
20. Lambropoulos, P., Maragakis, P., & Zhang, J. (1998). Two-electron atoms in strong fields. Physics Reports, 305(5), 203–293.
21. Liu, S., & Langenaeker, W. (2003). Hund's multiplicity rule: a unified interpretation. Theoretical Chemistry Accounts, 110, 338–344.
22. Lloyd, J. P., & Pake, G. E. (1954). Spin relaxation in free radical solutions exhibiting hyperfine structure. Physical Review, 94(3), 579.
23. Maki, K. (1966). Effect of Pauli paramagnetism on magnetic properties of high-field superconductors. Physical Review, 148(1), 362.
24. Matsen, F. A. (1964). Spin-free quantum chemistry. Advances in Quantum Chemistry, 1, 59–114.
25. Morgan III, J. D., & Kutzelnigg, W. (1993). Hund's rules, the alternating rule, and

symmetry holes. The Journal of Physical Chemistry, 97(10), 2425–2434.

26. Moroni, E. G., Kresse, G., Hafner, J., & Furthmüller, J. (1997). Ultrasoft pseudopotentials applied to magnetic Fe, Co, and Ni: From atoms to solids. Physical Review B, 56(24), 15629.
27. Mulders, A. M., Gubbens, P. C. M., Gasser, U., Baines, C., & Buschow, K. H. J. (1998). Two types of magnetism in the magnetic superconductor TmNi 2 B 2 C related to the degree of carbon-site occupancy. Physical Review B, 57(17), 10320.
28. Murray, É. D., Lee, K., & Langreth, D. C. (2009). Investigation of exchange energy density functional accuracy for interacting molecules. Journal of Chemical Theory and Computation, 5(10), 2754–2762.
29. Natoli, S. N., & McMillin, D. R. (2018). Demonstrating Hund's Rule in Action by Exploring the Magnetic Properties of Metal Complexes with 3d n and 4f n Configurations. Journal of Chemical Education, 95(1), 126–130.
30. Néel, L. (1955). Some theoretical aspects of rock-magnetism. Advances in physics, 4(14), 191–243.
31. Nesterenko, V. O., Kleinig, W., Reinhard, P. G., Iudice, N. L., Cruz, F. D. S., & Marinelli, J. R. (2003). Orbital magnetism in axially deformed sodium clusters: From scissors mode to dia-para magnetic anisotropy. The European Physical Journal D-Atomic, Molecular, Optical and Plasma Physics, 27, 43–53.
32. Nguyen, T. N., Lee, P. A., & Zur Loye, H. C. (1996). Design of a random quantum spin chain paramagnet: Sr3CuPt0. 5Ir0. 5O6. Science, 271(5248), 489–491.
33. Novoa, J. J., Deumal, M., & Jornet-Somoza, J. (2011). Calculation of microscopic exchange interactions and modelling of macroscopic magnetic properties in molecule-based magnets. Chemical Society Reviews, 40(6), 3182–3212.
34. Pyper, N. C., & Grant, I. P. (1977). On the interpretation of Hund's rules in atomic spectra. Journal of Physics B: Atomic and Molecular Physics, 10(10), 1803.
35. Ramirez, A. P., Cheong, S. W., & Kaplan, M. L. (1994). Specific heat of defects in Haldane systems Y 2 BaNiO 5 and NENP: Absence of free spin-1/2 excitations. Physical review letters, 72(19), 3108.
36. Sahni, V., Gruenebaum, J., & Perdew, J. P. (1982). Study of the density-gradient expansion for the exchange energy. Physical Review B, 26(8), 4371.
37. Schwinger, J. (1948). On quantum-electrodynamics and the magnetic moment of the electron. Physical Review, 73(4), 416.
38. Smirnov, B. M., & Chibisov, M. I. (1966). The breaking up of atomic particles by an electric field and by electron collisions. Sov. Phys. JETP, 22(585), 23.
39. Sondheimer, E. H., & Wilson, A. H. (1951). The diamagnetism of free electrons. Proceedings of the Royal Society of London. Series A. Mathematical and Physical Sciences, 210(1101), 173–190.
40. Szilva, A., Kvashnin, Y., Stepanov, E. A., Nordström, L., Eriksson, O., Lichtenstein, A. I., & Katsnelson, M. I. (2023). Quantitative theory of magnetic interactions in solids. Reviews of Modern Physics, 95(3), 035004.

41. Szmytkowski, R. (2002). Larmor diamagnetism and Van Vleck paramagnetism in relativistic quantum theory: The Gordon decomposition approach. Physical Review A, 65(3), 032112.
42. Van Vleck, J. H. (1928). On dielectric constants and magnetic susceptibilities in the new quantum mechanics Part III—application to dia-and paramagnetism. Physical Review, 31(4), 587.
43. Wilsdon, B. H. (1925). CXIV. The magnetic properties of atoms and molecules. The London, Edinburgh, and Dublin Philosophical Magazine and Journal of Science, 49(294), 1145–1164.
44. Zellinger, W., Moser, B. A., Grubinger, T., Lughofer, E., Natschläger, T., & Saminger-Platz, S. (2019). Robust unsupervised domain adaptation for neural networks via moment alignment. Information Sciences, 483, 174–191.
45. Zhao, M. M., Yang, Y., Du, N., Zhu, Y. Y., Ren, P., & Yen, F. (2024). Determining the chemical composition of diamagnetic mixed solids via measurements of their magnetic susceptibility. Journal of Materials Chemistry C, 12(16), 5877–5885.

CHAPTER

7

The Electronic Band Structure of Solids

LEARNING OBJECTIVES

After studying this chapter, you will be able to:

- Understand the concept of energy bands in crystals using reciprocal lattice and Bloch wave functions;
- Understand band formation in crystals via gradual introduction of periodic potential;
- Describe energy band splitting and Bragg reflection effects in crystals;
- Understand the Tight-Binding Approximation for electronic band structure in crystals;
- Understand the concept and calculation of density of states in electronic band structures.

KEY TERMS FROM THIS CHAPTER

Band structure	Bloch waves
Brillouin zone	Density of states
Energy bands	Periodic potential
Reciprocal lattice	Schrodinger equation
Tight-binding approximation	Van Hove singularities

UNIT INTRODUCTION

While the square-well potential and one-electron approximation are oversimplifications, the free-electron-gas model has been able to explain electron activity in crystals. Therefore, it is unrealistic to rely on this model to provide an explanation for the crucial optical and electronic properties of semiconductors. When analyzing the creation of a solid, it is crucial to acknowledge that the progressive aggregation of initially independent atoms leads to the emergence of the distinct energy levels of the separated atoms in the solid's characteristics (Parratt, 1959).

Discrete energy levels are necessary to account for the distinct resonance-like patterns found in the optical spectra of solids. The free-electron-gas concept does not provide an explanation for such phenomena. Moreover, this model fails to provide any insight into the characteristics of semiconductors and insulators. In order to achieve additional advancements, it is crucial to consider that the electrical states in solids are organized into what are known as bands. These come from unbound atom states. Current approximation ignores any departures from perfect periodicity, whether static lattice disturbances or dynamic atom oscillations. Assuming an endlessly stretched potential implies disregarding any surface effects (Tancogne et al., 2017).

One of the fundamental topics in solid-state physics is the study of electronic band structures in solids. The basic ideas and approximations that explain how electrons behave in crystalline materials are covered in this unit (Friedel, 1954). Topics covered include the Tight-Binding Approximation, which treats electrons as localized close to atomic sites, and the Nearly Free-Electron Approximation, which sees electrons as quasi-free inside a periodic lattice. Analyzing the distribution of accessible electronic states requires an understanding of state density (Morales et al., 2017).

7.1. GENERAL SYMMETRY PROPERTIES

Assuming periodicity in the potential V(r), the challenge at hand is to solve the time-independent Schrodinger equation for a single electron:

$$H\psi(r) = \left[-\frac{\hbar^2}{2m}\nabla^2 + V(r)\right]\psi(r) = E\psi(r) \tag{7.1}$$

where:

$$V(r) = V(r + r_n); r_n = n_1a_1 + n_2a_2 + n_3a_3 \tag{7.2}$$

The three-dimensional periodic lattice's translation vector is denoted by the sign . In other words, is composed of multiples , ,) of the real-space lattice's three basis vectors (, ,):

Given the periodicity of the lattice, the potential V(r) can be conveniently expanded using a Fourier series:

$$V(r) = \sum_G V_G e^{iG\cdot r} \tag{7.3}$$

The G vector must be at least a reciprocal lattice vector.

$$G = hg_1 + kg_2 + \lg_3,\ h,k,l \text{ int}\, egers. \tag{7.4}$$

In the context of one-dimensional systems, there exists a well-established relationship between G and G=h 2 π/a. One possible way to expand the wave function ψ(r) is through a general plane-wave expansion (Trench, 2004).

$$\psi(r) = \sum_k C_k e^{ik\cdot r} \tag{7.5}$$

In this case, k stands for a reciprocal space point that complies with the periodic boundary conditions. The following outcome can be obtained by changing the expansions (7.3) and (7.5) in the Schrondinger equation (7.1):

$$\sum_{k} \frac{\hbar^2 k^2}{2m} C_k e^{ik \cdot r} + \sum_{k'G} C_{k'} V_G e^{i(k'+G) \cdot r} = E \sum_{k} C_k e^{ik \cdot r} . \tag{7.6}$$

By renaming the summation indices, the expression is transformed into the following:

$$\sum_{k} e^{ik \cdot r} \left[\left(\frac{\hbar^2 k^2}{2m} - E \right) C_k + \sum_{G} V_G C_{k-G} \right] = 0 . \tag{7.7}$$

This means that the equation inside the brackets is independent of position vector v, must disappear for each k, which means that it must equal zero (Barros et al., 2006). This criterion applies to all position vectors v.

$$\left(\frac{\hbar^2 k^2}{2m} - E \right) C_k + \sum_{G} V_G C_{k-G} = 0 . \tag{7.8}$$

This set of equations represents the Schrodinger equation (7.1) in reciprocal space. It connects the expansion coefficients of ψ (r) (7.5) that have different k-values by a reciprocal lattice vector G. Therefore, is connected to, and so on (Bracken & Grundland, 2001).

As a result, the original problem can be divided into N different problems, where N is the number of unit cells. In the unit cell of the reciprocal lattice, each of these problems is related to a certain k-vector (Zabrodsky et al., 1992). Each set of equations can be expressed as a combination of plane waves with wave vectors that differ only by reciprocal lattice vectors G.

The eigenvalues E of the Schrodinger equation (7.1) can be categorized as =E (k), and the corresponding wave function is associated with .

$$\psi_k (r) = \sum_{G} C_{k-G} e^{i(k-G) \cdot r} \tag{7.9}$$

or

$$\psi_k(r) = \sum_{G} C_{k-G} e^{-iG \cdot r} e^{ik \cdot r} = u_k (r) e^{ik \cdot r} . \tag{7.10a}$$

The function displayed here exhibits lattice periodicity as it is a Fourier series defined over reciprocal lattice points G. For periodic boundary conditions, the wave vector k can have the following values

$$\begin{aligned} k_x &= 0, \ \pm 2\pi/L, \ \pm 4\pi/L, \ldots, \ \pm 2\pi n_x/L \\ k_y &= 0, \ \pm 2\pi/L, \ \pm 4\pi/L, \ldots, \ \pm 2\pi n_y/L \\ k_z &= 0, \ \pm 2\pi/L, \ \pm 4\pi/L, \ldots, \ \pm 2\pi n_z/L \end{aligned} \tag{7.10b}$$

(L=the crystal's macroscopic dimension) produces the accurate quantum numbers, , , or , , , which can be used to index the quantum states and energy eigenvalues. We showed that a modulated plane wave can express the periodic potential's one-electron Schrodinger equation solution:

$$\psi_k(r) = u_k(r)\, e^{i k \cdot r} \tag{7.10c}$$

featuring a modulation function that exhibits the same periodicity as the lattice. Bloch's theorem refers to this outcome, and the wave functions provided in equations (7.10 a–d) are referred to as Bloch states or Bloch waves of an electron (see Figure 7.1) (Urmantsev, 1986).

$$u_k(r) = u_k(r + r_n) \tag{7.10d}$$

Because of the lattice potential's precise regularity, additional conclusions can be drawn directly from the characteristics of the Bloch states.

Using reciprocal lattice vectors , and Bloch wave (7.10 a) tgeneral representation, we may deduce that:

$$\begin{aligned}\psi_{k+G}(r) &= \sum_{G'} C_{k+G-G'}\, e^{-iG' \cdot r}\, e^{i(k+G)\cdot r} \\ &= \left(\sum_{G''} C_{k-G''}\, e^{-iG'' \cdot r}\right) e^{ik\cdot r} = \psi_k(r)\,,\end{aligned} \tag{7.11a}$$

$$\psi_{k+G}(r) = \psi_k(r)\,. \tag{7.11b}$$

Therefore, if a reciprocal lattice vector separates two wave vectors, then the two Bloch waves are equivalent. Schrodinger's formula (7.1):

$$\mathscr{H}\, \psi_k = E(k)\, \psi_k \tag{7.12}$$

and that for the same issue displaced by G:

$$\mathscr{H}\, \psi_{k+G} = E(k+G)\psi_{k+G} \tag{7.13}$$

When combined with 7.11 b, the result is obtained:

$$\mathscr{H}\, \psi_k = E(k+G)\psi_k\,. \tag{7.14}$$

Comparing (7.14) with (7.12) we notice that:

$$E(k) = E(k+G)\,. \tag{7.15}$$

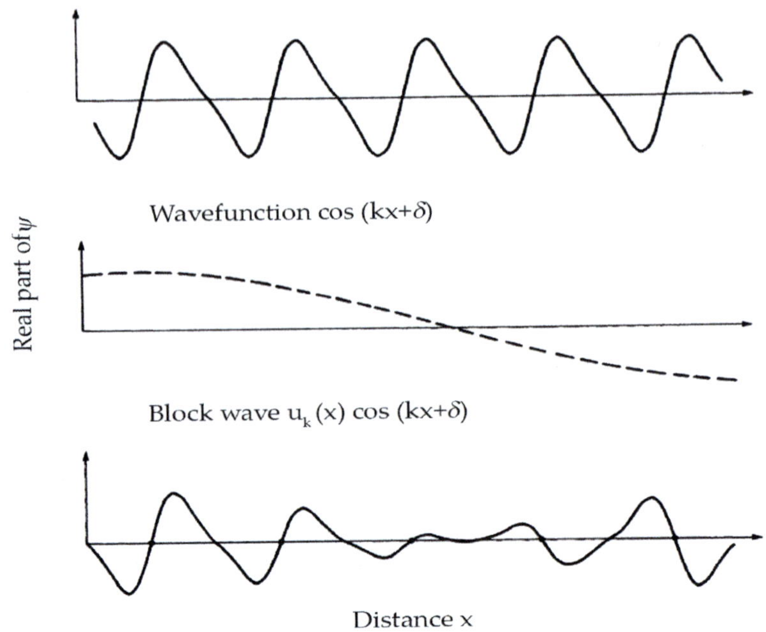

Figure 7.1. Illustration of an example of how to create a Bloch wave (r) = (from a plane wave and a lattice periodic function (r) with p-type bonding character.

Source: Huns Luth, Creative Commons License.

As a result, periodical behavior is observed for the energy eigenvalues E (k) with respect to the quantum numbers k, that is, BW vectors (Metiu et al., 1974). Periodic functions of the wave vector in reciprocal k-space can be used to characterize the energy surfaces of one-electron states under a periodic potential. This is similar to how phonons have dispersion surfaces in reciprocal q-space to describe their behavior.

These energy surfaces come together to form the crystal's electrical band structure. Since E (k) and (r) are periodic functions in reciprocal space, knowing these functions is sufficient for k-values in the first Brillouin zone. By extending the functions periodically, we can determine their values across the entire k-space (Yang, 1996).

7.2. THE NEARLY FREE-ELECTRON APPROXIMATION

In order to grasp a fundamental idea of electronic bands, it is highly beneficial to examine a scenario where a periodic potential becomes extremely small. It is hypothesized that the periodic potential initially has a value of zero and is then gradually activated (Northrup et al., 1989).

When free electrons are exposed to a different potential, their energy states undergo changes.

Specifically, how does the energy parabola . Change? It is crucial to take into account the periodicity's symmetry requirements even in cases where the potential is zero. These requirements play a crucial role, even for the smallest non-vanishing potential. Equation (7.15) clarifies that the requirement of periodicity does not restrict electron states to a single parabola in k-space. Instead, these states can be located on parabolas that are shifted by any G-vector (Oliveira & Sturm, 1980).

$$E(k) = E(k+G) = \frac{\hbar^2}{2m} | k + G |^2 \tag{7.16}$$

Figure 7.2 shows this for the one-dimensional example . Only the first Brillouin zone needs to be used to depict E k's behavior because it is periodic in k-space. To accomplish this, just move the relevant portion of parabola by necessary multiple of G=2 π/a. We refer to this process as "Reduction to the fiust Buillouin zone."

In a three-dimensional setting, equation (7.16) now incorporates G contributions in all three coordinate directions, leading to an even more complex band at zero potential. Figure 7.3 displays the E (k) curves along in the first Brillouin zone of a virtually zero potential simple cubic lattice.

We may now analyze the effect of a small but very tiny potential with reference to Figures 7.2 and 7.3. A degeneracy in the energy values is present in the one-dimensional scenario shown in Figure 7.2 at the limits of the first Brillouin zone, that is, at +G/2 = π/a and −G/2 = −π/a, where two parabolas overlap (Bayliss, 1952).

Remember

The Nearly Free-Electron Approximation simplifies analysis of electrons in crystals by considering them to be slightly influenced by the periodic potential of the lattice. This approach focuses on their behavior as quasi-free particles.

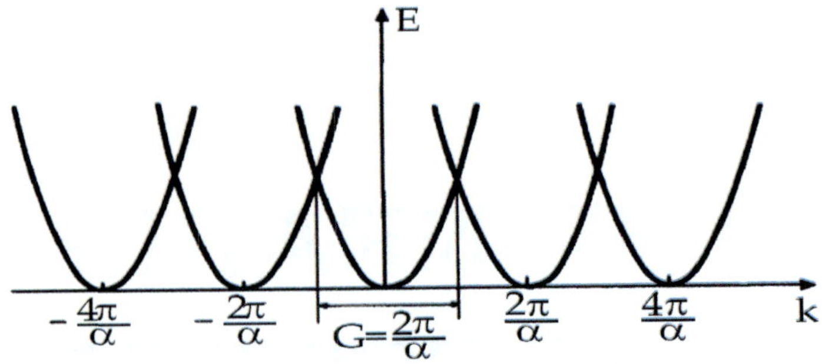

Figure 7.2. Illustration of free electron's one-dimensional parabolic energy curves are consistently expanded in reciprocal space. In real space, periodicity is a. The E(k) dependency represents the periodic lattice with diminishing potential, also known as the empty lattice.

Source: Grosee P., Creative Commons License.

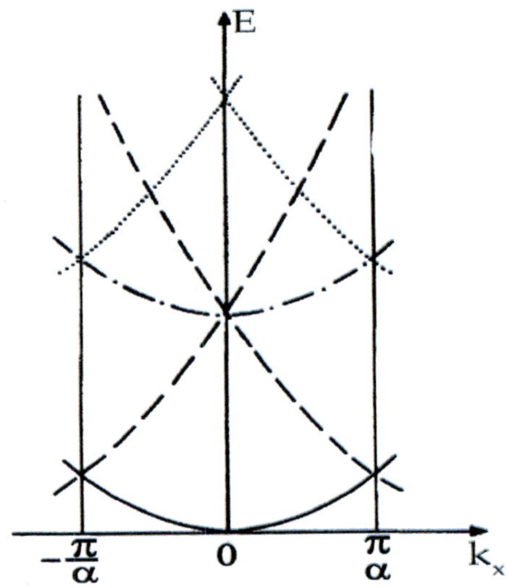

Figure 7.3. Illustration of a free electron gas band structure in the first Brillouin zone is shown along the direction in a primitive cubic lattice with a lattice constant .

Source: W. Shockley, Creative Commons License.

When considering the electron's state with these k-values, it is important to note that it is a combination of multiple plane waves (Khazaei et al., 2016). In the zeroth-order approximation, where the potential is minimal, these waves are:

$$e^{iGx/2} \quad \text{and} \quad e^{i[(G/2)-G]x} = e^{-iGx/2} . \tag{7.17}$$

Equation (7.8) indicates that waves with G-values greater than 2 π/a should be taken into consideration. On dividing (7.8) by it can be observed that becomes significantly large when and are both approximately equal to , and the coefficient exhibits a similar absolute magnitude as . Specifically, the scenario described in (7.17) includes the two plane waves located at the edges of the zone. Consequently, one can rule out the influence of other reciprocal lattice vectors in an initial study. Thus, the following arrangement of equations is appropriate for a perturbation estimate of the impact of a small potential:

$$\psi_+ \sim \left(e^{iGx/2} + e^{-iGx/2}\right) \sim \cos \pi \frac{x}{a}, \qquad (7.18a)$$

$$\psi_- \sim \left(e^{iGx/2} - e^{-iGx/2}\right) \sim \sin \pi \frac{x}{a}. \qquad (7.18b)$$

These are stationary waves that have points of zero amplitude at specific locations in space. Within the framework of diffraction from periodic structures, it has been established that these immobile waves can be characterized as a fusion of an entering wave and a Bragg-reflected wave that exhibits motion in the opposite direction. The probability densities, corresponding to ψ+ and ψ−, are depicted in Fig. 7.4 together with a qualitative sketch of the potential (Rösler & Brauer, 1981).

$$\varrho_+ = \psi_+^* \psi_+ \sim \cos^2 \pi \frac{x}{a}, \qquad (7.19a)$$

$$\varrho_- = \psi_-^* \psi_- \sim \sin^2 \pi \frac{x}{a}, \qquad (7.19b)$$

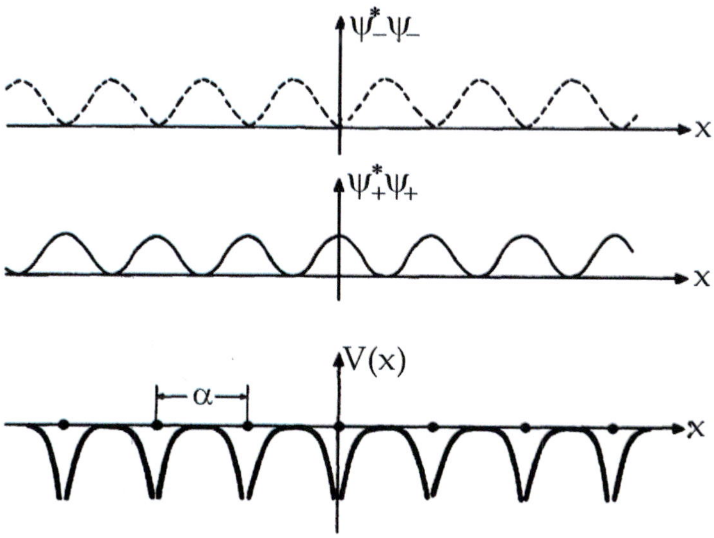

Figure 7.4. Illustration of the qualitative depiction of the electron's potential energy V(x) in a one-dimensional lattice. The ion cores positions are denoted by points that are separated by a distance of , which is a lattice constant.

Source: R Courthus, Creative Commons License.

An electron's charge density is highest in the state ψ+ at the site of the positive cores and lowest between them; in the state ψ–, it is highest in the space between the cores. It is significant to observe that, when comparing the travelling plane wave, the potential energy of ψ+ is smaller than that of the solution situated farther away from the zone boundary. On the other hand, ⊠– has more energy on the energy parabola in the zero potential situation than a free electron (Hede & Carbotte, 1972).

The deviation from the anticipated energy parabola of unbound electrons is indicated by the fluctuation in energy levels near the zone boundary (Figure 7.5).

Once you have a deeper understanding of the topic from this thorough discussion, it becomes easier to perform a formal calculation of the energy gap or band splitting as shown in Figure 7.5. By utilizing a reciprocal lattice vector, we can derive the representation of the general Schrodinger equation in k-space (7.8).

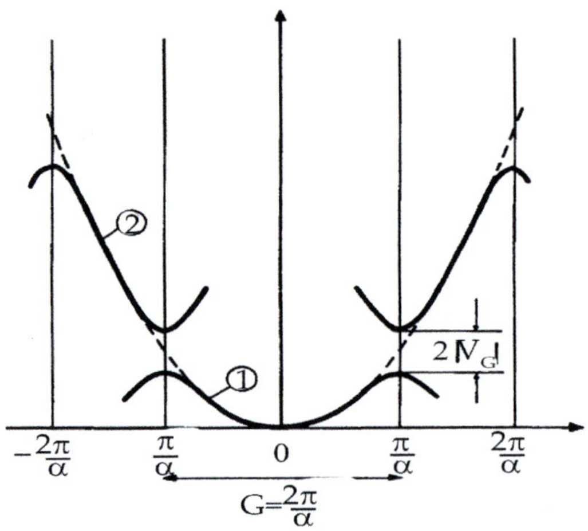

Figure 7.5. Illustration of division of the free electron's energy parabola.

Source: L.P Howard, Creative Commons License.

$$\left(E-\frac{\hbar^2}{2m}|k-G|^2\right)C_{k-G}=\sum_{G'}V_{G'}C_{k-G-G'}=\sum_{G'}V_{G'-G}C_{k-G'},\quad \text{i.e.} \tag{7.20a}$$

$$C_{k-G}=\frac{\sum_{G'}V_{G'-G}C_{k-G'}}{E-\frac{\hbar^2}{2m}|k-G|^2}. \tag{7.20b}$$

When dealing with minor disturbances, we can make an initial estimate for the calculation of by equating the desired eigenvalue E to the energy of the unbound electron. In addition, in this initial approximation, the coefficients that are the largest are the ones that we are interested in (Wang et al., 1999). This suggests that the greatest deviation from free-electron behavior will be expected as tbrahe denominator in equation (7.20 b) approaches zero, particularly for:

$$k^2 \simeq |k - G|^2 . \tag{7.21}$$

This and the Bragg condition are comparable. When the Bragg condition is met, the periodic potential has the most significant impact on the energy surface of the free electron, specifically for the k-vectors near the initial boundary of the Brillouin zone. It can be deduced from equation (7.20 b) that in addition to , the coefficient holds equal significance (Zhao et al., 2009). Therefore, in the given system (7.20 a), for this particular approximation, we only need to focus on two relationships :

$$\begin{aligned} &\left(E - \frac{\hbar^2}{2m}k^2\right) C_{\boldsymbol{k}} - V_{\boldsymbol{G}} C_{\boldsymbol{k}-\boldsymbol{G}} = 0 \\ &\left(E - \frac{\hbar^2}{2m}|\boldsymbol{k} - \boldsymbol{G}|^2\right) C_{\boldsymbol{k}-\boldsymbol{G}} - V_{-\boldsymbol{G}} C_{\boldsymbol{k}} = 0 . \end{aligned} \tag{7.22}$$

Hence, we have the secular equation for the energy value.

$$\begin{vmatrix} \left(\frac{\hbar^2}{2m}k^2 - E\right) & V_{\boldsymbol{G}} \\ V_{-\boldsymbol{G}} & \left(\frac{\hbar^2}{2m}\right)|\boldsymbol{k} - \boldsymbol{G}|^2 - E \end{vmatrix} = 0 . \tag{7.23}$$

When considering free electrons energy as , it is possible to express the two solutions to this secular equation in a certain manner (Hjelmberg et al., 1977).

Therefore, the two waves with and have equal contributions at the zone boundary, as indicated in equation 7.21 (Kevorkian & Li, 1988). Additionally, when , the energy gap assumes a specified value:

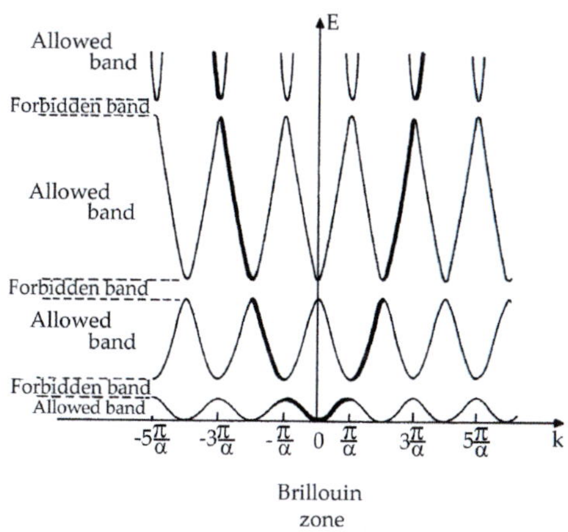

Figure 7.6. Illustration of the energy dispersion curves E(k) represent the relationship between energy and wavevector k for a one-dimensional lattice with a lattice constant a, using the extended zone technique.

Source: Wei Chen, Creative Commons License.

$$\Delta E = E_+ - E_- = 2|V_G|, \tag{7.25}$$

i.e., twice the Gth Fourier component of the potential.

The shape of the two energy surfaces near the boundary of the zone, which are separated by this gap, can be characterized by equation (7.24), where , = . Figure 7.5 demonstrates this phenomenon in the one-dimensional scenario, namely around the zero limit at k=G/2.

The relationship between the periodic band structure, which has energy gaps coming from the actual potential, and the parabolic energy distribution of free electrons is shown in Figures 7.6 and 7.5. These figures clearly show the relationship for one-dimensional scenario (Reininger et al., 1983).

7.3. THE TIGHT-BINDING APPROXIMATION

A free atom's electrons that reside in its energetically lower core levels show a high degree of spatial localization. When an atom takes part in the development of a crystal, it maintains its strong localization. Therefore, it seems that the electrical structure of a solid cannot be adequately explained by quasi-free electrons.

Combining atomic Eigen functions in a linear superposition is one way to characterize the crystal electrons, considering the similarities between these core electrons and the free atom. The Linear Combination of Atomic Orbitals (LCAO) approach is often used in the field of chemical bonding to explain the presence of electronic bands in solids (Reich et al., 2002).

When constructing the problem, it is assumed that the unbound atoms that make up the crystal have solutions to the Schrodinger equation are known.

$$H_A(r-r_n)\varphi_i(r-r_n)=E_i\varphi_i(r-r_n) \tag{7.26}$$

The Hamiltonian represents the energy operator for an unbound atom located at the lattice location. The equation a3 represents the position vector as a linear combination of vectors a3 where are coefficients. The function represents the wave function for an electron in the atomic energy level Ei. It is expected that the crystal is composed of individual atoms, and the Hamiltonian for an electron (assuming only one electron) can be expressed as the sum of the potentials from all the atoms (Smerzi & Trombettoni, 2003).

$$H=H_A+v=-\frac{\hbar^2}{2m}\Delta+V_A(r-r_n)+v(r-r_n) \tag{7.27}$$

The influence of atoms near the highly concentrated electron of interest, near, can be accounted for by a perturbation to the unbound atom's potential . Consequently, this disruption can be shown as:

$$v(r-r_n)=\sum_{m\neq n}V_A(r-r_m) \tag{7.28}$$

That is, the electron's position € is determined by summing the potentials of all atoms except the one at (Figure 7.7). Currently, we are in search of solutions to the Schrodinger equation (Mookherjea & Yariv, 2001).

$$\mathcal{H}\psi_{k}(r) = E(k)\psi_{k}(r) \,, \tag{7.29}$$

The symbol represents the Hamiltonian of equation (7.27) for the electron in the crystal. The functions ψk (v) are Bloch waves and possess certain general characteristics (Koskinen & Mäkinen, 2009).

By multiplying (7.29) with the complex conjugate of and integrating across the given range of , one may easily derive:

$$E(k) = \frac{\langle \psi_k | \mathcal{H} | \psi_k \rangle}{\langle \psi_k | \psi_k \rangle} \,, \tag{7.30}$$

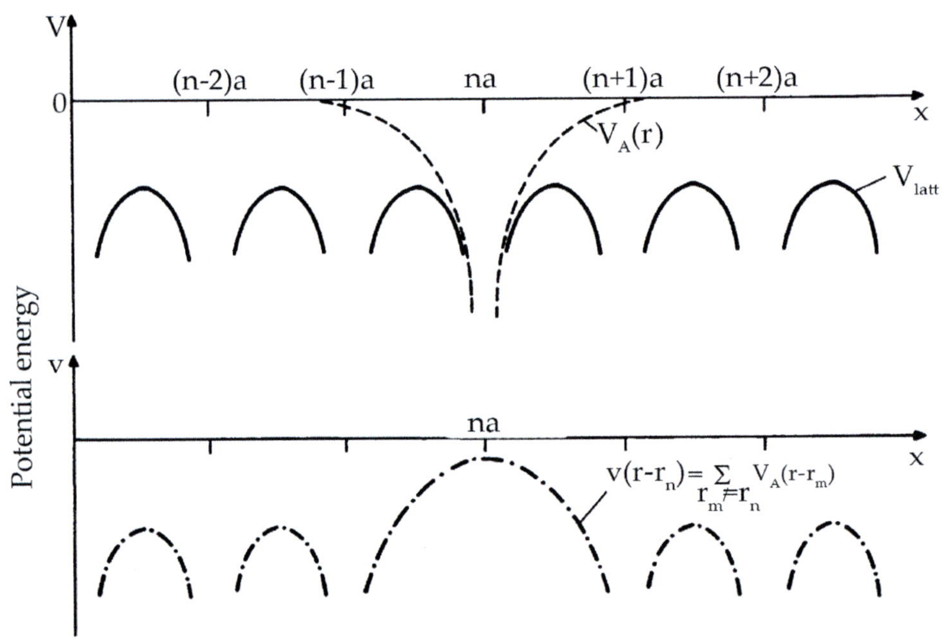

Figure 7.7. Illustration of the potential cross section along the x-direction used in the tight-binding approximation.

Source: R.W Powell, Creative Commons License.

When analyzing the current situation, in order to determine the energy states E(k) of crystal electrons derived from the energy level of free atoms, we make the assumption that ψk can be effectively approximated by a linear combination of atomic Eigen functions

$$\psi_k \approx \Phi_k = \sum_n a_n \varphi_i(\mathbf{r} - \mathbf{r}_n) = \sum_n e^{i\mathbf{k}\cdot\mathbf{r}_n} \varphi_i(\mathbf{r} - \mathbf{r}_n) . \tag{7.31}$$

The condition that Φk must be a Bloch wave determines the expansion coefficients. These coefficients (Del Re, 1963), denoted as . It is evident that Φk in (7.31) exhibits all the necessary characteristics of Bloch waves, e.g.:

$$\Phi_{k+G} = \sum_n e^{i\mathbf{k}\cdot\mathbf{r}_n} e^{i\mathbf{G}\cdot\mathbf{r}_n} \varphi_i(\mathbf{r} - \mathbf{r}_n) = \Phi_k . \tag{7.32}$$

By substituting the trial function from equation (7.31) into equation (7.30), we can now do an estimated computation of E (k). This will result in a modified denominator in equation (7.30).

$$\langle \Phi_k | \Phi_k \rangle = \sum_{n,m} e^{i\mathbf{k}\cdot(\mathbf{r}_n - \mathbf{r}_m)} \int \varphi_i^*(\mathbf{r} - \mathbf{r}_m) \varphi_i(\mathbf{r} - \mathbf{r}_n) d\mathbf{r} . \tag{7.33}$$

The function only has significant values around for an electron that is sufficiently confined (Boykin & Klimeck, 2005). Therefore, in order to simplify the equation (7.33), we only consider terms when n is equal to m, resulting in:

$$\langle \Phi_k | \Phi_k \rangle \simeq \sum_n \int \varphi_i^*(\mathbf{r} - \mathbf{r}_n) \varphi_i(\mathbf{r} - \mathbf{r}_n) d\mathbf{r} = N , \tag{7.34}$$

where: N is the crystal's atomic number.

We write (7.26) knowing the free atom solutions.

$$E(\mathbf{k}) \approx \frac{1}{N} \sum_{n,m} e^{i\mathbf{k}\cdot(\mathbf{r}_n - \mathbf{r}_m)} \int \varphi_i^*(\mathbf{r} - \mathbf{r}_m)[E_i + v(\mathbf{r} - \mathbf{r}_n)] \varphi_i(\mathbf{r} - \mathbf{r}_n) d\mathbf{r} \tag{7.35}$$

where: stands for the isolated atom's energy eigenvalue. Once again, we have ignored the interaction between neighboring particles in the term involving and have only looked at terms where n = m. Keep in mind that the overlap is limited to the closest neighbor when analyzing the perturbation ,.

When the atomic state has spherical symmetry, the result can be easily represented using two quantities (Seifert, 2007).

$$A = -\int \varphi_i^*(\mathbf{r} - \mathbf{r}_n) v(\mathbf{r} - \mathbf{r}_n) \varphi_i(\mathbf{r} - \mathbf{r}_n) d\mathbf{r} , \tag{7.36 a}$$

$$B = -\int \varphi_i^*(\mathbf{r} - \mathbf{r}_m) v(\mathbf{r} - \mathbf{r}_n) \varphi_i(\mathbf{r} - \mathbf{r}_n) d\mathbf{r} \tag{7.36 b}$$

and reads:

$$E(\boldsymbol{k}) \approx E_i - A - B \sum_m e^{i\boldsymbol{k}\cdot(\boldsymbol{r}_n - \boldsymbol{r}_m)} . \tag{7.37}$$

When considering the sum over 3, it is important to only include values where represents a nearest neighbor of .

Since is negative in this particular case, A is positive. When using equation (7.37) on a primitive cubic lattice with (Fang et al., 2015).

$$\boldsymbol{r}_n - \boldsymbol{r}_m = (\pm a, 0, 0) ; \ (0, \pm a, 0) ; \ (0, 0, \pm a) ,$$

gives, for an atomic s-state:

$$E(\boldsymbol{k}) \approx E_i - A - 2B(\cos k_x a + \cos k_y a + \cos k_z a) . \tag{7.38}$$

When atoms are combined to create a crystal, the individual atomic energy levels transform into electronic bands. In the case of a crystal with a primitive cubic lattice, the center of gravity of the band is shifted by an amount A compared to the original energy level , while the width of the band is proportional to B (Manousakis, 1991). This scenario is shown in Figure 7.8.

Here are the key outcomes that can be summarized:

- The cosine terms vary from +1 to −1, therefore 12 is the breadth of the energy band. The cosine terms can be stretched for small k-values so that one obtains . At the ⊠-point (the center of the first Brillouin zone at k= 0). The dependence on k is in line with the findings of the quasi-free-electron approximation (Paxton, 2009);

$$E(\boldsymbol{k}) = E_i - A - 6B + Ba^2k^2 , \tag{7.39}$$

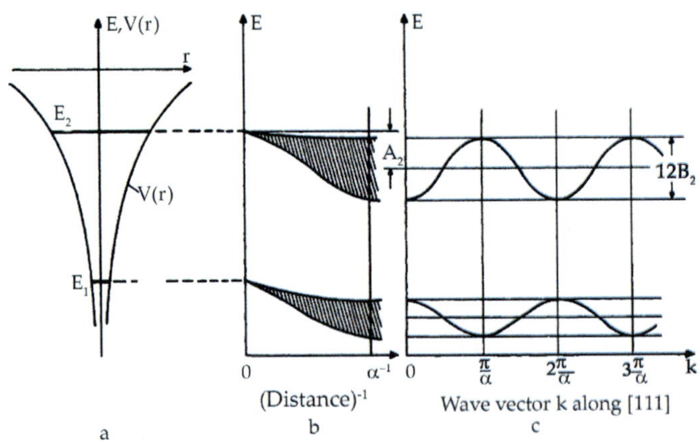

Figure 7.8. Illustration of an example of tight-binding computations' qualitative outcome for a primitive cubic lattice.

Source: H. Ibach, Creative Commons License.

- According to equation (Equation 7.36b), the energy range of the band increases when there is a larger overlap between the wave functions of surrounding atoms. Bands arising from more confined states have shorter widths compared to bands emerging from less constrained atomic states with more extended wave functions (Mitra, 1969);
- The occupation of the one-electron band states is determined, within the bounds of the current one-electron approximation, by assigning two electrons from each atom to each band, beginning with the lowest energy band and continuing until all electrons have been assigned. One electron can be occupied twice, according to the Pauli principle, as there are two different spin states (Barišić, 1972).

A free atom's atomic energy level splits into N states when it interacts with other (N−1) atoms in a crystal with a primitive cubic lattice composed of N atoms, each of which represents a primitive unit cell. Consequently, the matching quasi-continuous band is produced. This means that a total of two N electrons can fit in this band (Sankey & Niklewski, 1989). Analyzing this problem from the viewpoint of the quasi-free-electron model leads to the same result. Every electron state has a volume of , in k-space, where V is the macroscopic crystallographic volume. , is the volume of the first Brillouin zone; the unit cell's volume is denoted by . N states make up the portion of the band that is contained in the first Brillouin zone; the ratio of this zone's volume to the entire volume is denoted by V/. Electrons can occupy a total of 2N states when accounting for the two spin states (Boykin et al., 2007). The qualitative illustration demonstrated the emergence of a band structure resulting from the discrete energy levels of individual atoms when they combine to form a crystal. In the case of sodium, the atomic 3s and 3p orbitals result in overlapping bands in the crystal structure at the equilibrium atomic spacing . The occupied energy levels of atomic sodium are . The 3s-band of the crystal receives only one electron per unit cell from the atom's 3s-level. Nevertheless, the 3s-band has the capacity to accommodate two electrons per unit cell. Therefore, even in the absence of the 3 s-3 p overlap, the 3 s band of sodium would still only be partially occupied (Mattheiss, 1970).

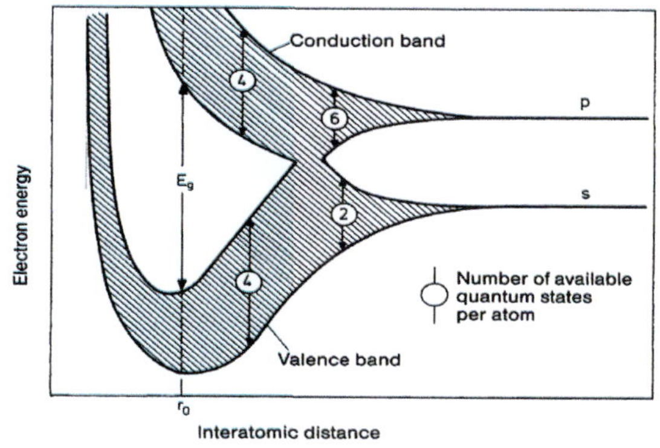

Figure 7.9. The energy band behavior of the tetrahedral semiconductors diamond €, Si, and Ge is shown schematically as a function of atomic spacing.

Source: G. Simmons, Creative Commons License.

The electronic configuration of isolated carbon atoms is commonly recognized as . However, in the diamond crystal, the development of the hybrid leads to a modification of the sand p-levels. This change causes the hybrid band to split into two bands. Each of these bands may contain four electrons, including spin (Figure 7.9).

The lower portion of the band is occupied by the four electrons from the atomic 2s- and 2p-states, while upper portion is unoccupied (Mönch, 1996). An energy gap, known as , exists between the two sub bands, and it is forbidden for electrons to occupy this region. The band structure observed in Figure 7.9 cannot be obtained using the simple approach described in this specific situation. This calculation requires the utilization of sophisticated methodologies that rely on state-of-the-art computing resources (Gert et al., 2015).

7.4. THE DENSITY OF STATES

The density of states is used by both the electronic states and the thermal properties of the phonon system to explain phenomena like the energy of the electron system. The quantity of electron states within each energy interval dE can often be used to understand the spectra when examining specific electronic excitation techniques, such as non-angle-resolved photoemission spectroscopy, that studies electron momentum in all directions (Alexander & Orbach, 1982).

After obtaining the energy surfaces E(k) of the band structure, the density of states can be determined by integrating across an energy shell in k-space, using a method similar to the one given in equation (5.4).

$$dZ = \frac{V}{(2\pi)^3} \int_{E}^{E+dE} dk, \tag{7.40}$$

Here, Vrepresents the density of states in k-space. By dividing the volume element dk into an area element on the energy surface and a perpendicular component dk⊥, the equation , can be applied (Figure 5.1) (Jacob et al., 2010). By utilizing the equation d perpendicular, we can deduce the subsequent outcome:

$$D(E)dE = \frac{1}{(2\pi)^3}\left(\int_{E(k)=const} \frac{df_E}{|\, grad_k E(k) \,|} \right) dE \tag{7.41}$$

The density of states D € is expressed in terms of the actual volume V of the crystal to derive a quantity that is particular to the crystal. It is important to note that, because of spin degeneracy, each state has the capacity to hold two electrons (Singh et al., 2012).

The sites in k-space where the gradient of E with respect to becomes zero, showing that the energy surfaces are flat, determine the fundamental configuration in the function D (E). Van Hove singularities or critical points are the names given to these particular places (White & Mintmire, 1998).

In three dimensions, the singularity of D (E) around these critical points is avoided because when expanding E k about the extremum (), the inverse of the magnitude of the gradient of E −1 exhibits a singularity (Carminati et al., 2015).

Therefore, when we integrate the E (k) surface (7.41), we obtain a linear dependence on k. The states density around a critical point in three dimensions exhibits the specific pattern depicted in Figure 7.10. Band structures in one-dimensional systems, often employed to analyze organic semiconductors, display a divergence in the density of states at critical sites, while the overall integral of the density of states remains finite. (Refer to the linear chain lattice vibration density of states analysis) (Rathore et al., 2003).

Did you know?

The concept of Density of States (DOS) in solid-state physics not only informs about the availability of electronic states but also plays a crucial role in understanding the optical and thermal properties of materials.

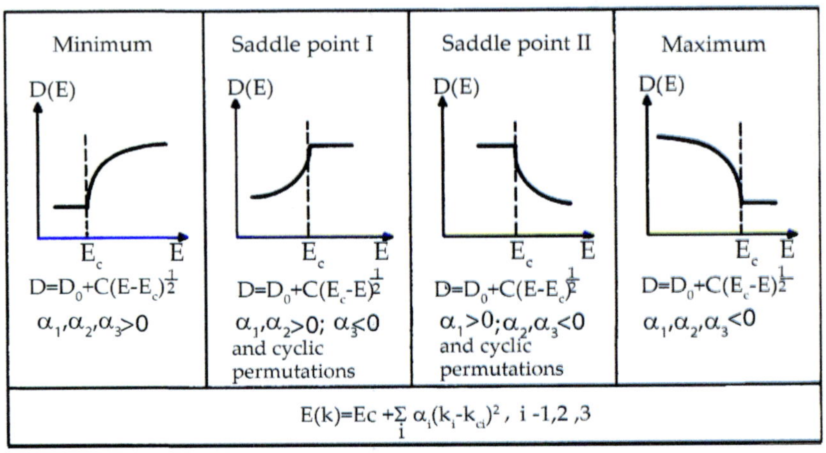

Figure 7.10. Illustration of density states D (E) at the four potential critical points types in three dimensions.

Source: G. Dolling, Creative Commons License.

Theoretical densities of states can be determined by integrating across the first Brillouin zone in k-space using a computed band structure. These densities can then be compared to experimental data, such as those acquired via photoemission spectroscopy (Panel V). Therefore, they serve as a crucial link between the estimated band structures and the experimental data. The density of states in the integration over k-space is mostly determined by the critical points (Chicanne et al., 2002).

Given that critical points typically arise along lines or at sites of significant symmetry in k-space, this provides another rationale for representing band structures along lines of high symmetry, such as ΓK, ΓX, ΓL, and so on. In the intermediate regions, the contributions are so little that they can be considered insignificant. Basic mathematical interpolation can sometimes determine the band structure in those places (Heremans et al., 2008). An effective demonstration of the correlation between a computed band structure and the matching density of states can be shown by examining the case of germanium, a semiconductor, as illustrated in Figure 7.11. Flat sections of the E k curves along high-symmetry directions are strongly associated to important contributions, especially the highest points in the density of states. At low temperatures, there is a significant disparity in energy between completely filled valence band states and empty conduction band states (Mao et al., 2001). The restricted band gap for germanium is approximately 0.7 electron volts (eV) in width.

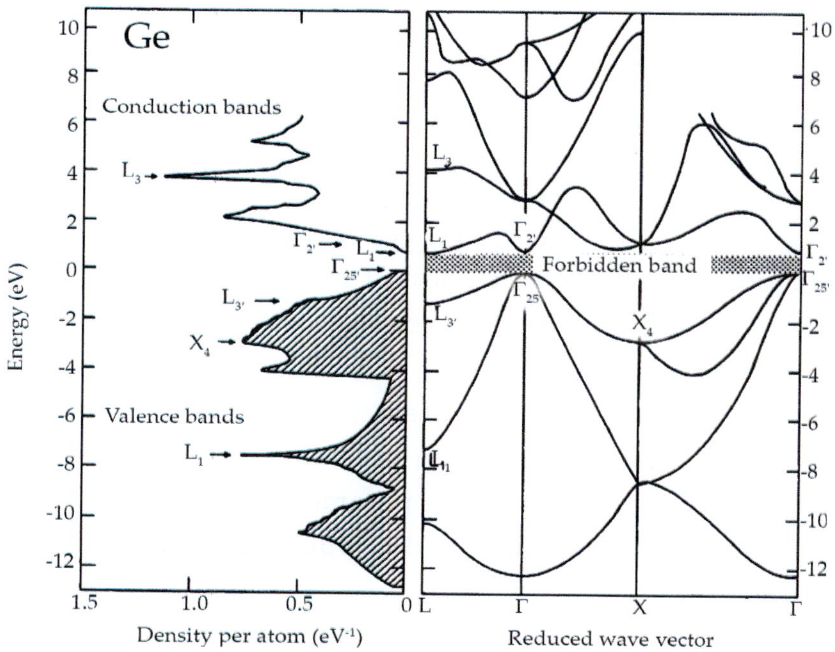

Figure 7.11. Illustration of germanium theoretically derived band structure E (k).

Source: M Born, Creative Commons License.

ACTIVITY 7.1

Objective: To explore electronic band structures using an online simulation tool.

Instructions:

1. Simulation Exploration:
   - Access the online simulation tool for electronic band structures.
   - Explore different types of materials (metals, insulators, semiconductors) and observe how band structures vary.
   - Adjust parameters such as lattice spacing, number of atoms, or electron interactions to see their impact on band structures.
   - Analyze the patterns and features of the band structures observed.
2. Discussion:
   - Discuss findings with classmates:
   - What patterns did you notice in the band structures of different materials?
   - How do changes in simulation parameters affect the band structures?
   - Relate observations to real-world applications or phenomena.

SUMMARY

- The chapter explores fundamental ideas of solid-state physics that are crucial for comprehending the behavior of electrons in crystalline materials. It delves into the symmetrical features that arise from the periodic arrangement of atoms in a crystal lattice, which have a substantial impact on electronic states. The Nearly Free-Electron Approximation is a technique used to approximate electrons as quasi-free particles within a crystal. This simplifies the investigation of electronic band structures.
- On the other hand, the Tight-Binding Approximation views electrons as being highly concentrated around atomic sites, where they mostly interact with neighboring atoms to create energy bands. The chapter also discusses the idea of the Density of States, which characterizes the allocation of energy states accessible to electrons in a material.

REVIEW QUESTIONS

1. Compare nearly free-electron and tight-binding approximations in solid-state physics.
2. Explain how crystal lattice arrangement affects electronic state symmetry. Provide examples.
3. Define density of states (DOS) in solid-state physics and its relevance to electronic properties.
4. Assess advantages and limitations of nearly free-electron and tight-binding approximations for different materials' band structures.
5. Discuss contributions of electronic band structure concepts to material conductivity and optical behavior, citing examples.
6. Contrast band structures of metals, insulators, and semiconductors using chapter principles.

MULTIPLE CHOICE QUESTIONS

1. Which approximation method treats electrons as nearly free within a crystal lattice?

a. Tight-Binding Approximation

b. Density Functional Theory

c. Nearly Free-Electron Approximation

d. `Hubbard Model

2. What does the Density of States describe in solid-state physics?

a. The distribution of atomic densities within a crystal lattice

b. The energy levels available to electrons within a material

c. The mass density of a solid material
d. The probability distribution of electron spins

3. **Which property of crystal lattices strongly influences the symmetry of electronic states?**
   a. Atomic radius
   b. Packing density
   c. Periodic arrangement of atoms
   d. Electronic charge distribution
4. **In the Tight-Binding Approximation, how are electrons modeled within a crystal?**
   a. As nearly free particles
   b. As localized near atomic sites
   c. As interacting via Coulomb forces
   d. As moving in a uniform potential
5. **Which theoretical framework is crucial for understanding the electronic properties of materials?**
   a. Quantum Mechanics
   b. Classical Mechanics
   c. General Relativity
   d. Solid-State Physics

Answers to Multiple Questions

1. (c); 2. (b); 3. (c); 4. (b); 5. (d)

REFERENCES

1. Alexander, S., & Orbach, R. (1982). Density of states on fractals:«fractons». Journal de Physique Lettres, 43(17), 625–631.
2. Barišiⓧ, S. (1972). Self-consistent electron-phonon coupling in the tight-binding approximation. II. Physical Review B, 5(3), 941.
3. Barros, E. B., Jorio, A., Samsonidze, G. G., Capaz, R. B., Souza Filho, A. G., Mendes Filho, J., ... & Dresselhaus, M. S. (2006). Review on the symmetry-related properties of carbon nanotubes. Physics reports, 431(6), 261–302.
4. Bayliss, N. S. (1952). The free-electron approximation for conjugated compounds. Quarterly Reviews, Chemical Society, 6(4), 319–339.
5. Boykin, T. B., & Klimeck, G. (2005). Practical application of zone-folding concepts in tight-binding calculations. Physical Review B—Condensed Matter and Materials Physics, 71(11), 115215.

6. Boykin, T. B., Kharche, N., Klimeck, G., & Korkusinski, M. (2007). Approximate bandstructures of semiconductor alloys from tight-binding supercell calculations. Journal of Physics: Condensed Matter, 19(3), 036203.
7. Bracken, P., & Grundland, A. M. (2001). Symmetry properties and explicit solutions of the generalized Weierstrass system. Journal of Mathematical Physics, 42(3), 1250–1282.
8. Carminati, R., Cazé, A., Cao, D., Peragut, F., Krachmalnicoff, V., Pierrat, R., & De Wilde, Y. (2015). Electromagnetic density of states in complex plasmonic systems. Surface Science Reports, 70(1), 1–41.
9. Chicanne, C., David, T., Quidant, R., Weeber, J. C., Lacroute, Y., Bourillot, E., ... & Girard, C. (2002). Imaging the local density of states of optical corrals. Physical review letters, 88(9), 097402.
10. Del Re, G. (1963). Hybridization and localization in the tight-binding approximation. Theoretica chimica acta, 1(2), 188–197.
11. Fang, S., Kuate Defo, R., Shirodkar, S. N., Lieu, S., Tritsaris, G. A., & Kaxiras, E. (2015). Ab initio tight-binding Hamiltonian for transition metal dichalcogenides. Physical Review B, 92(20), 205108.
12. Friedel, J. (1954). Electronic structure of primary solid solutions in metals. Advances in Physics, 3(12), 446–507.
13. Gert, A. V., Nestoklon, M. O., & Yassievich, I. N. (2015). Band structure of silicene in the tight binding approximation. Journal of Experimental and Theoretical Physics, 121, 115–121.
14. Hede, B. B. J., & Carbotte, J. P. (1972). Positron annihilation in a nearly-free electron band. Journal of Physics and Chemistry of Solids, 33(3), 727–735.
15. Heremans, J. P., Jovovic, V., Toberer, E. S., Saramat, A., Kurosaki, K., Charoenphakdee, A., ... & Snyder, G. J. (2008). Enhancement of thermoelectric efficiency in PbTe by distortion of the electronic density of states. Science, 321(5888), 554–557.
16. Hjelmberg, H., Gunnarsson, O., & Lundqvist, B. I. (1977). Theoretical studies of atomic adsorption on nearly-free-electron-metal surfaces. Surface Science, 68, 158–166.
17. Jacob, Z., Kim, J. Y., Naik, G. V., Boltasseva, A., Narimanov, E. E., & Shalaev, V. M. (2010). Engineering photonic density of states using metamaterials. Applied physics B, 100, 215–218.
18. Kevorkian, J., & Li, Y. P. (1988). Explicit Approximations for Strictly Nonlinear Oscillators with Slowly Varying Parameters with Applications to Free-Electron Lasers. Studies in Applied Mathematics, 78(2), 111–165.
19. Khazaei, M., Ranjbar, A., Ghorbani-Asl, M., Arai, M., Sasaki, T., Liang, Y., & Yunoki, S. (2016). Nearly free electron states in MXenes. Physical Review B, 93(20), 205125.
20. Koskinen, P., & Mäkinen, V. (2009). Density-functional tight-binding for beginners. Computational Materials Science, 47(1), 237–253.
21. Manousakis, E. (1991). Electronic structure of C 60 within the tight-binding approximation. Physical Review B, 44(19), 10991.

22. Mao, H. K., Xu, J., Struzhkin, V. V., Shu, J., Hemley, R. J., Sturhahn, W., ... & Wortmann, G. (2001). Phonon density of states of iron up to 153 gigapascals. Science, 292(5518), 914–916.

23. Mattheiss, L. F. (1970). Crystal-Field Effects in the Tight-Binding Approximation: Re O 3 and Perovskite Structures. Physical Review B, 2(10), 3918.

24. Metiu, H., Ross, J., Silbey, R., & George, T. F. (1974). On symmetry properties of reaction coordinates. The Journal of Chemical Physics, 61(8), 3200–3209.

25. Mitra, T. K. (1969). Electron-phonon interaction in the modified tight-binding approximation. Journal of Physics C: Solid-state physics, 2(1), 52.

26. Mönch, W. (1996). Empirical tight-binding calculation of the branch-point energy of the continuum of interface-induced gap states. Journal of applied physics, 80(9), 5076–5082.

27. Mookherjea, S., & Yariv, A. (2001). Optical pulse propagation in the tight-binding approximation. Optics Express, 9(2), 91–96.

28. Morales-García, Á., Valero, R., & Illas, F. (2017). An empirical, yet practical way to predict the band gap in solids by using density functional band structure calculations. The Journal of Physical Chemistry C, 121(34), 18862–18866.

29. Northrup, J. E., Hybertsen, M. S., & Louie, S. G. (1989). Quasiparticle excitation spectrum for nearly-free-electron metals. Physical Review B, 39(12), 8198.

30. Oliveira, L. E., & Sturm, K. (1980). High-frequency dielectric properties of covalent semiconductors within the nearly-free-electron approximation. II. The two-plasmon-band model. Physical Review B, 22(12), 6283.

31. Parratt, L. G. (1959). Electronic band structure of solids by x-ray spectroscopy. Reviews of Modern Physics, 31(3), 616.

32. Paxton, A. T. (2009). An introduction to the tight binding approximation–implementation by diagonalisation. NIC Series, 42, 145–176.

33. Rathore, N., Knotts IV, T. A., & de Pablo, J. J. (2003). Density of states simulations of proteins. The Journal of chemical physics, 118(9), 4285–4290.

34. Reich, S., Maultzsch, J., Thomsen, C., & Ordejon, P. (2002). Tight-binding description of graphene. Physical Review B, 66(3), 035412.

35. Reininger, R., Asaf, U., Steinberger, I. T., & Basak, S. (1983). Relationship between the energy V 0 of the quasi-free-electron and its mobility in fluid argon, krypton, and xenon. Physical Review B, 28(8), 4426.

36. Rösler, M., & Brauer, W. (1981). Theory of secondary electron emission. I. General theory for nearly-free-electron metals. physica status solidi (b), 104(1), 161–175.

37. Sankey, O. F., & Niklewski, D. J. (1989). Ab initio multicenter tight-binding model for molecular-dynamics simulations and other applications in covalent systems. Physical Review B, 40(6), 3979.

38. Seifert, G. (2007). Tight-binding density functional theory: an approximate Kohn–Sham DFT scheme. The Journal of Physical Chemistry A, 111(26), 5609–5613.

39. Singh, S., Chopra, M., & de Pablo, J. J. (2012). Density of states–based molecular simulations. Annual Review of Chemical and Biomolecular Engineering, 3(1), 369–394.

40. Smerzi, A., & Trombettoni, A. (2003). Nonlinear tight-binding approximation for Bose-Einstein condensates in a lattice. Physical Review A, 68(2), 023613.

41. Tancogne-Dejean, N., Mücke, O. D., Kärtner, F. X., & Rubio, A. (2017). Impact of the electronic band structure in high-harmonic generation spectra of solids. Physical review letters, 118(8), 087403.

42. Trench, W. F. (2004). Characterization and properties of matrices with generalized symmetry or skew symmetry. Linear algebra and its applications, 377, 207–218.

43. Urmantsev, Y. A. (1986). Symmetry of system and system of symmetry. Computers & Mathematics with Applications, 12(1–2), 379–405.

44. Wang, Y. A., Govind, N., & Carter, E. A. (1999). Erratum: Orbital-free kinetic-energy functionals for the nearly free electron gas [Phys. Rev. B 58, 13 465 (1998)]. Physical Review B, 60(24), 17162.

45. White, C. T., & Mintmire, J. W. (1998). Density of states reflects diameter in nanotubes. Nature, 394(6688), 29–30.

46. Yang, C. N. (1996). Symmetry and physics. Proceedings of the American Philosophical Society, 140(3), 267–288.

47. Zabrodsky, H., Peleg, S., & Avnir, D. (1992). Continuous symmetry measures. Journal of the American Chemical Society, 114(20), 7843–7851.

48. Zhao, J., Feng, M., Yang, J., & Petek, H. (2009). The superatom states of fullerenes and their hybridization into the nearly free electron bands of fullerites. ACS nano, 3(4), 853–864.

CHAPTER

8

Superconductivity

LEARNING OBJECTIVES

After studying this chapter, you will be able to:

- Know the Discovery and occurrence of superconductivity;
- Recognize the characteristics and factors influencing superconductivity;
- Understand the thermal and magnetic properties of superconductors, including changes in thermal conductivity and the Meissner effect;
- Describe the critical magnetic field and Meissner effect in superconductors;
- Understand the thermodynamic aspects of superconducting transitions, including entropy, latent heat, and specific heat characteristics;
- Understand the BCS theory of superconductivity, focusing on Cooper pairs and the energy gap parameter;
- Know key applications of superconductivity in diverse fields.

KEY TERMS FROM THIS CHAPTER

BCS theory	Electrical resistance
Energy gap	Kammerlingh Onnes
Liquid helium	Meissner effect
Mercury	Persistent current
Superconductivity	Transition temperature (Tc)

UNIT INTRODUCTION

Superconductivity is a phenomenon shown by some materials, in which electrical resistance completely disappears and magnetic fields are discharged from the substance. Any substance that displays these characteristics is considered a superconductor. In contrast to a typical metallic conductor, which experiences a steady decline in resistance as its temperature decreases, even approaching absolute zero, a superconductor has a distinct critical temperature at which its resistance suddenly becomes zero. A continuous flow of electric current can be maintained indefinitely in a loop of superconducting wire without the need for any external power source (Simon, 1997).

This chapter will begin by discussing the initial observation of superconductivity, followed by an explanation of several characteristics shown by superconductors. The topics covered encompass the exploration and presence of superconductivity (Tanigaki et al., 1991), characteristics such as complete absence of electrical resistance and ideal diamagnetism, theoretical frameworks such as the BCS theory, and practical applications spanning from power transmission to advanced technologies like SQUIDs and magnetic levitation (Putilin et al., 1993).

8.1. DISCOVERY OF SUPERCONDUCTIVITY

In 1908, Kammerlingh Onnes achieved the successful liquefaction of helium. This created a new low temperature range in which to research material properties. Onnes began researching metals' electrical resistance at temperatures close to liquid helium. Onnes made the discovery of superconductivity while researching mercury's electrical resistance (Van Delft, 2012).

Figure 8.1 displays the apparatus utilized by Onnes. Purified mercury is delivered into a vacuum in a series of U-tubes (A-B) with a diameter of 0.05 mm. The U-tubes are positioned within a cryostat that is specifically constructed to hold liquid helium. Within this cryostat, the mercury undergoes freezing and transforms into an elongated wire (Kamiya et al., 2018).

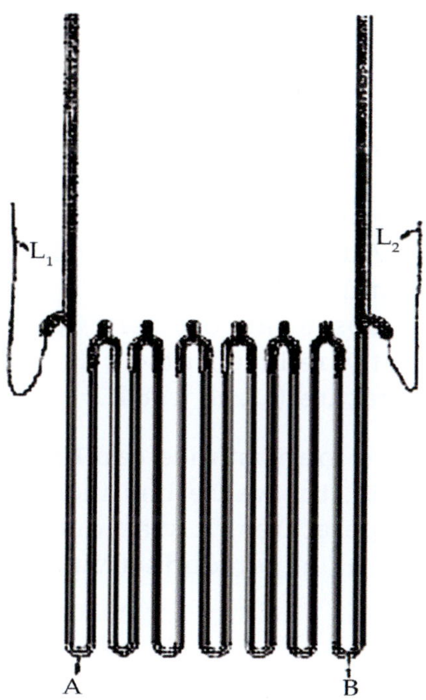

Figure 8.1. Illustration of the equipment utilized in the exploration of superconductivity by Kammerlingh Onnes.

Source: D.B. Sirdeshmukh, Creative Commons License.

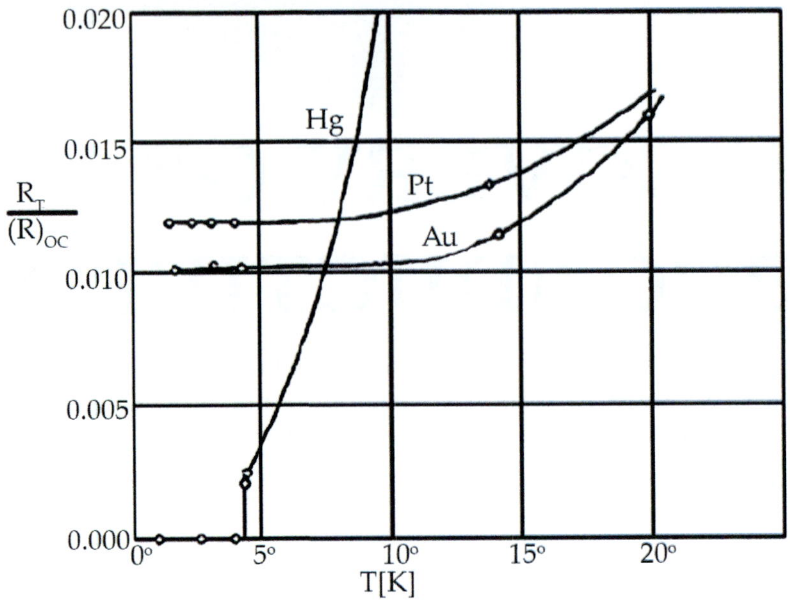

Figure 8.2. Illustration of temperature variation of mercury, gold, and platinum fractional resistance at very low temperature expressed as .

Source: C.S. Sunandana, Creative Commons License.

The platinum wires and are connected to the ends of the mercury wire in order to allow the flow of current and facilitate the measurement of potential difference.

Figure 8.2 displays the findings from Onnes' experiments on mercury and several other metals. The graph illustrates the variation in electrical resistance with changing temperatures. It is clear that when the temperature is dropped from about 20 K to *1.5 K, the resistance of Pt and Au shows a progressive decline. Mercury's resistance likewise gradually drops to around 5 K, though at a faster rate than that of Pt and Au. Nevertheless, it suddenly drops to almost zero at 4.2 K. According to Onnes, the resistance, which was significantly reduced at 4.2 K, dropped to an extremely small fraction of that at normal temperature (Dahl, 1984).

The sharp and abrupt decrease in resistance can be observed more clearly in Figure 8.3. When it comes to conductivity, the value has increased to a level that is a billion times higher than at typical temperatures. It was determined by Onnes that mercury has entered a novel state known as superconductivity (Eremets & Drozdov, 2016).

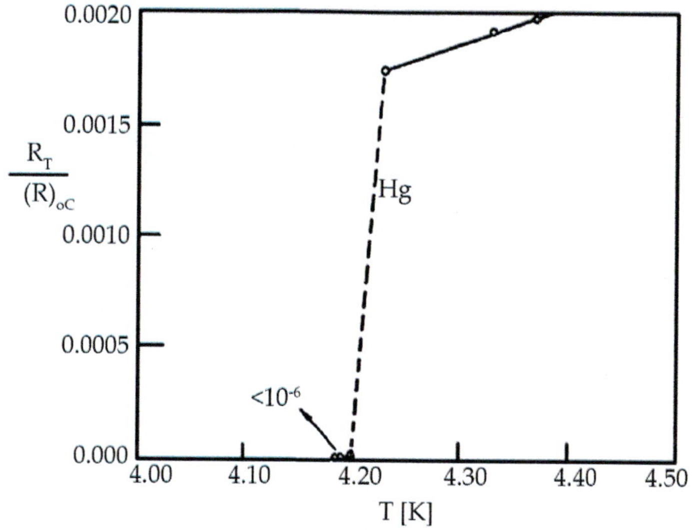

Figure 8.3. Illustration of abrupt decrease in the fractional resistance .

Source: L. Sirdeshmukh, Creative Commons License.

Onnes used the term "transition point" to refer to the temperature where superconductivity arises. The transition temperature is commonly known as such. When the temperature of a metal is below its critical temperature (), it exhibits superconductivity. However, when the temperature is above , the metal behaves like a normal metal (Larbalestier & Canfield, 2011).

8.2. OCCURRENCE

The discovery of superconductivity in mercury marked a significant milestone in scientific research, due to the pioneering work of Kammerlingh Onnes. During the course of his experiments, he made an interesting discovery – the occurrence of superconductivity in and Pb The discovery of superconductivity by Onnes sparked a wave of interest among various research groups and laboratories, leading to extensive studies in this field. As a result, the phenomenon of superconductivity has been observed in numerous other metals (Matthias, 1957). Here is a comprehensive list of metals that have superconductivity, along with their respective transition temperatures, as shown in Table 8.1. It is important to keep in mind that this list is not complete. Among the various elements that demonstrate superconductivity at normal pressures, niobium stands out with its remarkable transition temperature of 9.25 K (Emery et al., 2005).

Mathias put forward the following empirical rules regarding the manifestation of superconductivity:

- Superconductivity is observed in metallic elements with a range of 2 to 8 valence electrons per atom (R), as shown in Figure 8.4;
- Elements possessing an odd R exhibit elevated values, whereas elements with even R values have diminished values (Slocombe et al., 2015);

Table 8.1. Values offor Some Elements Which are Superconductors at Atmospheric Pressure

Element	**T_C (K)**
Niobium	9.25
Tantalum	4.47
Lead	7.19
Tin (grey)	3.72
Mercury	4.15
Tellurium	2.38
Indium	3.41
Cadmium	0.56
Aluminium	1.17
Gallium	1.08
Titanium	0.4

Source: L. Sirdeshmukh, Creative Commons License.

- Some elements that, when pure, are not superconductors turn into such when they are alloyed;

- Maximum has been recorded for compounds and alloys with R in the range of 5 and 7.

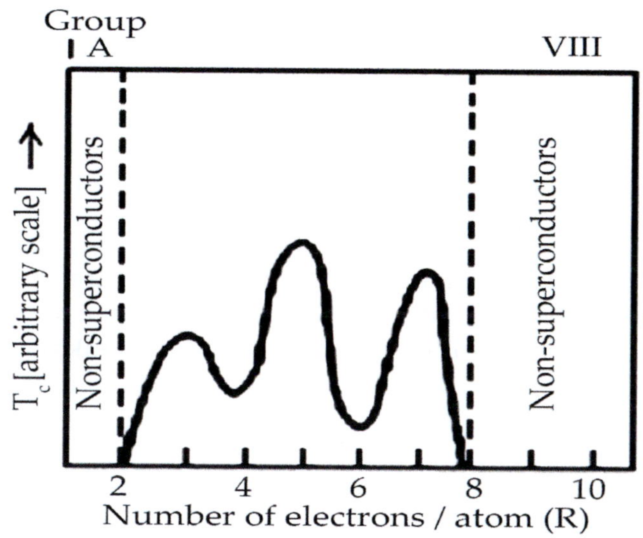

Figure 8.4. Plot of for several metals against the number of valence electrons (Z).

Source: K.G. Subhadra, Creative Commons License.

- A transition metal is not superconducting if its atomic radius (R) is smaller than 2. Furthermore, noble metals and monovalent metals do not display superconductivity under typical circumstances. Under normal air pressure circumstances, all rare earth elements except lanthanum do not display superconductivity, including ferromagnetic metals (Schmidt et al., 1995);
- Typically, superconductivity is observed in materials that have high resistance ρ. Based on empirical evidence, materials having a resistivity ratio exhibit superconductivity.

A few other observations are:

- The sample's purity level affects how sharply the transition temperature changes. It can occasionally happen at a temperature of just 0.001 K;
- Trace levels of paramagnetic elements have the ability to suppress superconductivity completely or to lower the transition temperature. As an example, molybdenum's superconductivity can be totally eliminated by just a few parts per million of iron. One atomic percent of gadolinium causes a sharp decrease in lanthanum's transition temperature, from 5.6 K to 0.6 K (Bednorz et al., 1987).

High pressure techniques have facilitated the examination of materials under elevated pressures. The previous piston-displacement techniques are limited to a maximum pressure of 25 GPa.

However, the development of diamond anvil methods has expanded the pressure range to 260 gigapascals (GPa). Various elements have been examined for potential superconductivity under these elevated pressures, and indeed, a significant number of them have demonstrated the presence of superconductivity (Cava et al., 1989).

Table 8.2 lists the elements, along with the pressure and , that are non-superconducting under normal pressures but turn into superconductors at greater pressures. These elements may have undergone distinct processes leading up to the advent of superconductivity.

The transition in crystal structure and the transition in electronic structure are two significant mechanisms (Finnemore et al., 2001).

One example of the first kind is bismuth. At normal pressure, it is non-superconducting in the trigonal structure; however, at 9 GPa, when it has a fcc structure, it becomes a superconductor with = 8.7 K. Under high pressure, the alkaline earth metals Ca, Sr, and Ba undergo an s-d electronic transition, resulting in superconductivity (Weihrich et al., 2011).

Table 8.2. values for Some Materials at Greater Pressure that are Superconducting

Element	Pressure (GPa)	T_C (K)
Si	12.0	8.5
Ba	20.0	5.0
Sb	36.0	8.5
Ca	150	4.0
Bi	9.0	8.7
Fe	21.0	2.0
Ge	11.5	5.4
Sr	50.0	4.0
Ce	5.0	1.7
Cs	8.0	1.7

Source: Herald I, Creative Commons License.

$\leq$ 10 K for the systems in Tables 8.1 and 8.2. A higher superconductor has been sought for in light of potential uses. Table 8.4 lists many superconducting compounds. For a few of them, > 10 K. , or the temperature between liquid hydrogen and liquid nitrogen, is 23 K for Nb3Ge, the highest of these (Abrikosov, 1957).

The field of superconductivity has been in existence for a century. Throughout this time frame, the phenomenon of superconductivity has been detected in a wide range of substances, including many natural elements, alloys, numerous inorganic compounds, as well as organic compounds and the double-stranded molecule of DNA. Thorough examination of all these systems is outside the limits of our focus (Liu et al., 2011).

In conclusion, it is worth noting the significant finding made by Bednorz and Muller on the observation of superconductivity in ceramic materials with the formula (M = Ba or Sr) at a critical temperature ≈ 40 K. The ceramic compound was quickly discovered to have a critical temperature ≈ 93 K, which is higher than the temperature of liquid nitrogen. We will address this feature in a subsequent section (Araujo-Moreira et al., 2000).

Remember

Superconductors exhibit zero electrical resistance below their critical temperature, enabling lossless electrical transmission.

8.3. PROPERTIES OF SUPERCONDUCTORS

When a metal transitions into a superconductor, certain properties, including thermal, magnetic, and thermodynamic properties, undergo significant changes. The presence or absence of a magnetic field determines the impact on thermodynamic properties during a transition. In this section, we will delve into the thermal and magnetic properties, as well as the thermodynamic properties (Samokhin, 2004) (Table 8.3).

Table 8.3. Summary of Information on Elemental Superconductors

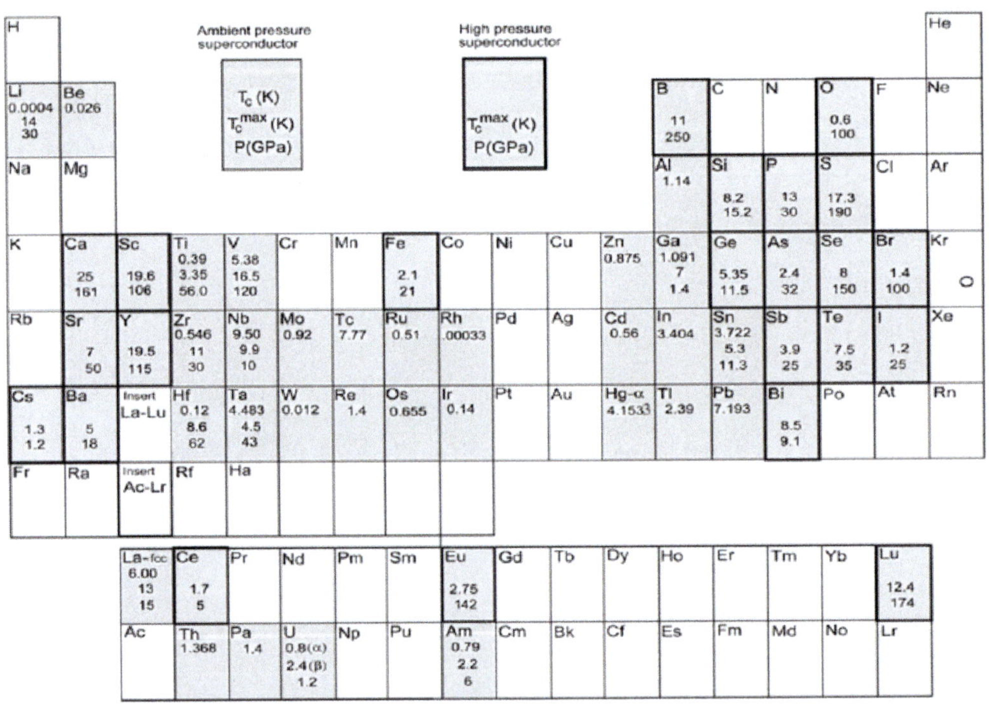

Source: Herald I, Creative Commons License.

Table 8.4. Values of for Some Superconducting Compounds

Compound	T_c [K]
Nb_3Ge	23.20
Nb_3Ga	20.30
Nb_3Sn	18.05
Nb_3Al	18.60
ZrN	13.70
NbN	17.30
InSb	1.90
NiBi	4.25

Source: Husten K, Creative Commons License.

8.3.1. Thermal Properties

8.3.1.1. Thermal Conductivity

Below the transition point, the thermal conductivity of tin experiences a significant alteration, gradually approaching zero. The magnitude of this change is almost directly proportional to the cube of the transition temperature. Figure 8.5 shows the variation of lead's thermal conductivity with temperature (Toberer et al., 2012). Curve N represents the heat conductivity in the normal state, while curve S represents the thermal conductivity in the superconducting state. As can be shown, the thermal conductivity (K) sharply decreases with decreasing temperature below Tc. Conversely, the normal state curve exhibits a smooth and constant decline as temperature rises (Carruthers, 1961).

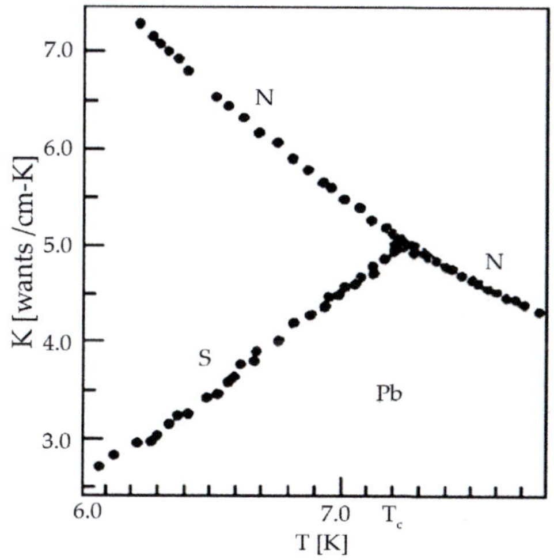

Figure 8.5. Illustration of the Pb's thermal conductivity in relation to temperature.

Source: T.E. Faber, Creative Commons License.

Good heat conductors are typically those that conduct electricity well. Superconductors have low thermal conductivity despite being excellent electrical conductors (Bristow, 2002). This suggests that a portion of electrons do not transfer heat (entropy) when a normal conductor transforms into a superconductor at . This is what makes superconducting electrons unique (Makinson, 1938,).

Did you know?

With hydrogen sulfide at extremely high pressures, the greatest critical temperature () yet observed in a superconductor is about 138 K (-135°C), suggesting that room temperature superconductivity may be possible in the future.

8.3.1.2. Thermoelectric Effect

The thermoelectric effects are absent from an electrical circuit consisting of two superconductors or one superconductor and one ordinary conductor. The lack of the Peltier effect suggests that there is no entropy carried by electrons in the superconducting phase. This suggests that the superconducting state's entropy is lower than the ordinary state's (Bergfield et al., 2010).

8.3.2. Magnetic Properties

8.3.2.1. Critical Field

Superconductivity can be eliminated by the existence of a strong magnetic field. A field can be produced by either an externally applied force or by an electric current flowing within the metal. The critical field, , is the minimum value of the applied field required to suppress superconductivity. The critical field varies with temperature (Werthamer et al., 1966). A parabola that is represented by the following equation can be used to approximate the variation of a critical field as a function of temperature, which is denoted by T:

$$H_c(T) = H_0\left[1 - \left(\frac{T}{T_c}\right)^2\right] \quad (8.1)$$

represents the critical magnetic field required to eliminate superconductivity at the temperature of absolute zero.

The critical field's temperature dependency for a typical superconductor is shown in Figure 8.6. Throughout the entire curve, the material is in a normal condition when it is above and in a superconducting state when it is below . Figure 8.7 illustrates the temperature dependence of for several specific elements (Konczykowski et al., 1991).

8.3.2.2. Meissner Effect

It was observed by Meissner and Ochsenfeld that the magnetic field application to a superconductor at a specific temperature , result in the complete expulsion of the magnetic field (Figure 8.8a). When the tin and lead samples were exposed to a magnetic field at a temperature lower than their transition temperature, the distribution of magnetic flux outside the samples was observed. Based on their experiments, it was observed that there are distinct differences in the magnetic behavior between a perfect conductor and a superconductor (Geim. 1998).

This phenomenon cannot be fully explained by considering only the fact that a superconductor is a substance with zero electrical resistance. Their research revealed that a superconductor, aside from being an excellent conductor, also exhibits perfect diamagnetism, meaning that the ratio of magnetic field (B) to magnetic field strength (H) is equal to zero. The induced magnetization in the specimen counteracts the applied field and completely neutralizes it. The occurrence is referred to as the Meissner effect. Arkadyev first demonstrated the Meissner effect with the famous 'suspended magnet' experiment (Figure 8.8b). When the magnet's magnetic field lines encounter resistance from the superconductor's surface, the magnet is elevated. The phenomenon is widely known as magnetic levitation (Bouchard et al., 2011). Detection of the Meissner effect with a diamond magnetometer.

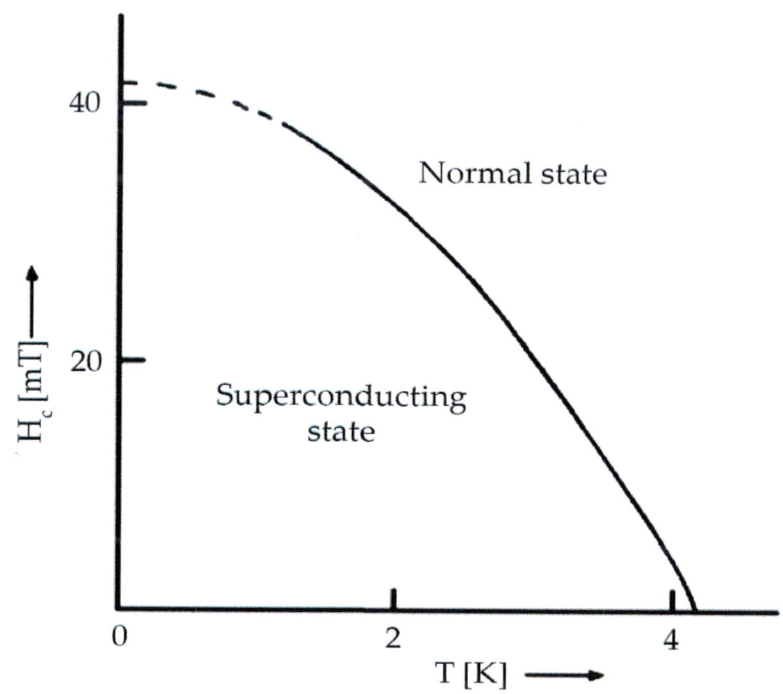

Figure 8.6. Schematic plot of the relationship between the critical magnetic field and temperature.

Source: J. Bardeen, Creative Commons License.

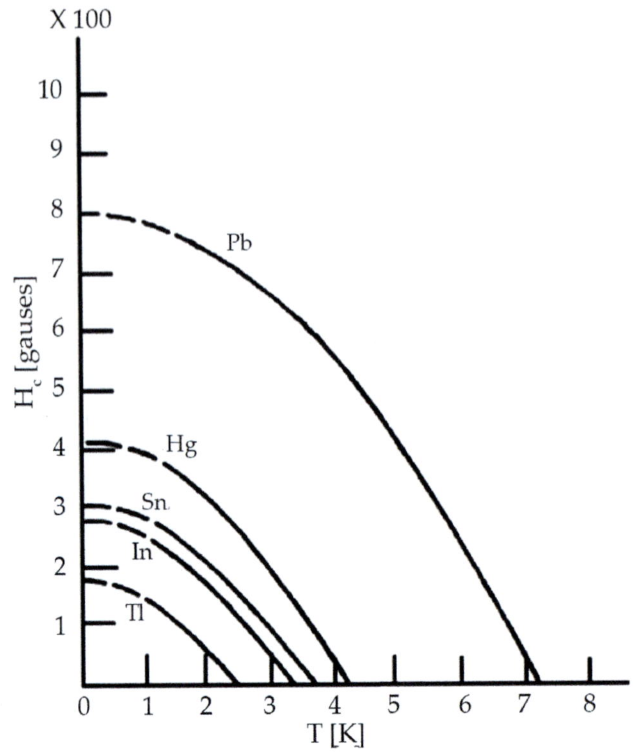

Figure 8.7. Illustration of variation of critical field temperature for certain superconducting components.

Source: J. Bardeen, Creative Commons License.

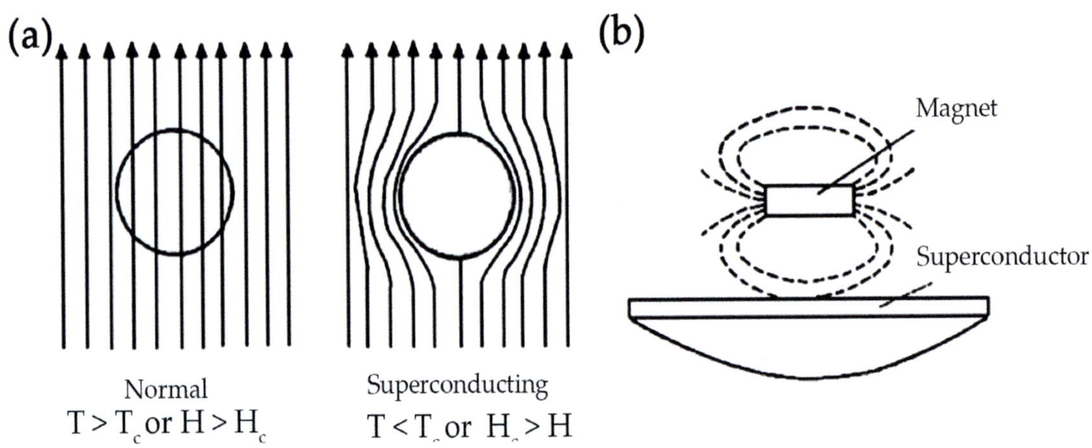

Figure 8.8. Illustration of (a) The Meissner effect and (b) Levitation principle.

Source: D.R. Tilley, Creative Commons License.

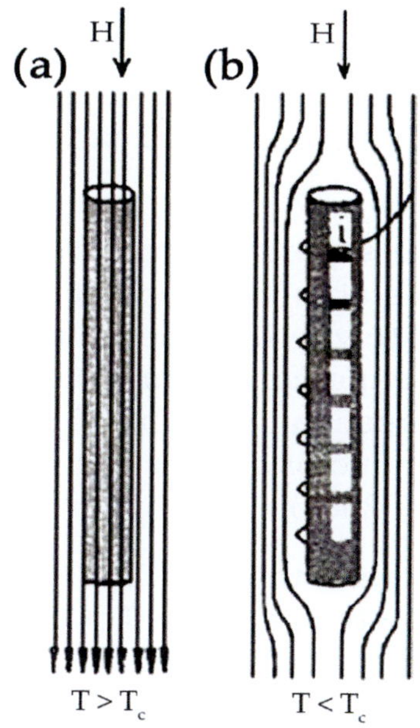

Figure 8.9. Illustration of flux distribution surrounding a superconducting rod at (a) T > and (b) T < ; observe the exclusion of flux caused by the induction current I.

Source: R.C. Jaklevic, Creative Commons License.

In Figure 8.9a, it is observed that a magnetic field should be applied parallel to the long axis of the long rod sample of a superconductor. The flux completely vanishes when the sample is cooled below the critical temperature. This is the result of the sample's induction current (Figure 8.9b) (Galambos et al., 2022).

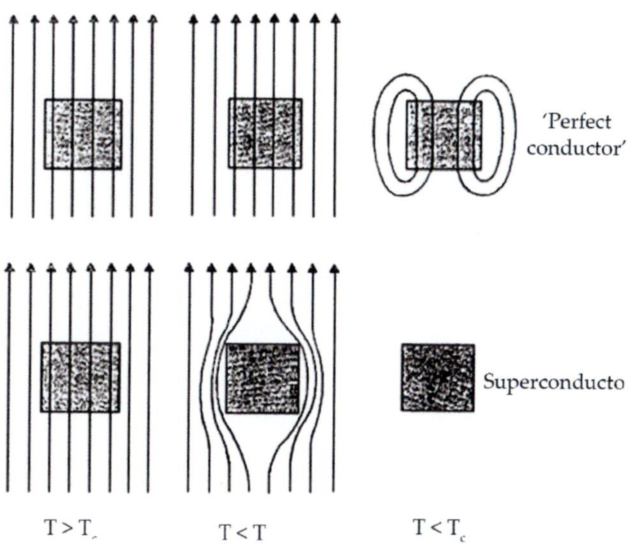

Figure 8.10. Illustration of flux distribution difference between hypothetical perfect conductor and a superconductor.

Source: S. Shapiro, Creative Commons License.

Now, using solid blocks as our example, we compare the flux arrangement in a "perfect conductor" with a super-conductor (Figure 8.10). First, let's look at a flawless conductor (first row). The presence of the field causes the sample to acquire induced current at T > and T < , but the induced field stays frozen in the sample when the field H = 0.

When the field is present, the superconductor in the second row is initially at T > (the usual state). Because the metal is superconducting, the flux is expelled when T < and the field is below . Ultimately, when the field terminates, an opposing surface current is induced, leaving the superconductor field devoid of current. It should be emphasized that whereas a superconductor's shift from a superconducting to a normal state is reversible, a normal conductor's transition from a "perfect conductor" to a normal state is not (Suzuki et al., 2005).

In this analysis, we will investigate the distribution of flux in a superconducting ring, as shown in Figure 8.11. When the temperature T > and the magnetic field is activated, the magnetic flux penetrates throughout the entire system. When the temperature of the sample is lowered to a value lower than the critical temperature (T <), and the magnetic field is still present, the magnetic flux is prevented from passing through the cross-section of the ring due to the induction current circulating in the loop.

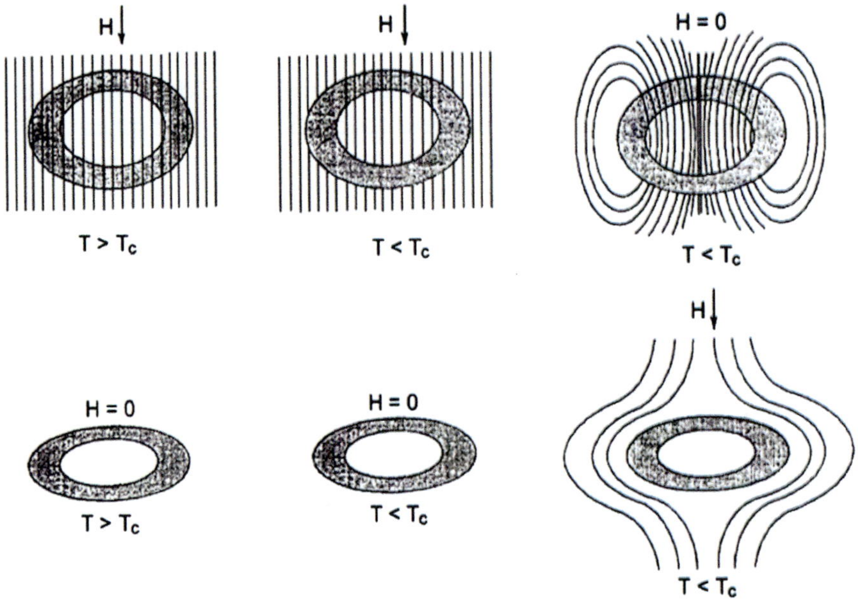

Figure 8.11. Distribution of flux around a superconducting ring with varying cooling and field application sequences.

Source: K. Kubo, Creative Commons License.

Ultimately, when the magnetic field is removed, the flux within the ring becomes confined as it is unable to permeate through the superconductor. The confined flux is sustained by a strong generated current. This electrical current continues to flow even after the magnetic field is removed due to the absence of resistance. The phenomenon being referred to is known as the persistent current (Hirsch & Marsiglio, 2021). Conversely, the ring in the second row is initially at T > and at zero field. After cooling to T < T_c, the field remains zero. Ultimately, the flux is excluded from both the ring and the hole when the ring is at T < and the field is turned on because the flux lines are unable to traverse the superconductor. In this particular scenario, unlike the one with the solid block, the arrangement of the flux is influenced by whether the magnetic field is applied prior to or after the sample is cooled. Ensuring the absence of magnetic flux within the sample is a vital aspect of the Meissner effect (Kozhevnikov et al., 2021).

Within the range of – cm, a magnetic field can enter a superconductor, unlike the electric field which is unable to do so (Bouchard et al., 2011).

8.4. THERMODYNAMICS OF SUPERCONDUCTING TRANSITION

Thermodynamic variables can be used to explain the phase transitions from liquid to vapour and from the normal to superconducting state. This will let us calculate latent heat, specific heat, and entropy at temperature and critical field. We analyze a type I superconductor with the full Meissner effect (, or zero flux B).

The equation provides the value of the Gibb's free energy G for each unit of volume.

$$G = U - TS - HM \quad 8.2$$

where: T is the temperature; S is the entropy; H is the magnetic field; M is the magnetization; and U is the internal energy. Since we are working with a magnetic system, –HM is used in place of pV in the conventional equation in this equation. Using from the second law of thermodynamics and differentiating (8.2), we have at constant temperature (Kumar et al., 1999).

$$dG = -MdH = \frac{H}{4\pi} dH \quad 8.3$$

In the superconducting state, the Gibb's free energy) for a field increase from zero to H is

$$\int_o^H dG = \int_o^H \frac{H}{4\pi} dH = \frac{H^2}{8\pi} \quad 8.4$$

or,

$$G_S(H) - G_S(0) = H^2 / 8\pi \quad 8.5$$

When the critical field, , is reached, the superconducting and normal states achieve a state of equilibrium at a specific temperature, T.

Hence:

$$G_n(T, H_c) = G_s(T, H_c) = G_s(0) + H_c^2/8\pi. \quad (8.6)$$

Figure 8.12a displays the change in free energy measured for aluminum as a function of temperature. It is evident that is greater than for temperatures . At the temperature of transition, the curves combine (Rao et al., 984).

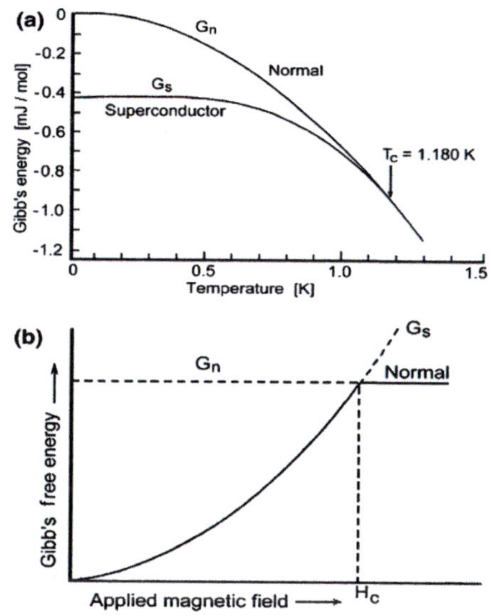

Figure 8.12. (a) and experimental values for Al at low temperature. (b) and free energies in relation to an applied magnetic field.

Source: G. Burns, Creative Commons License.

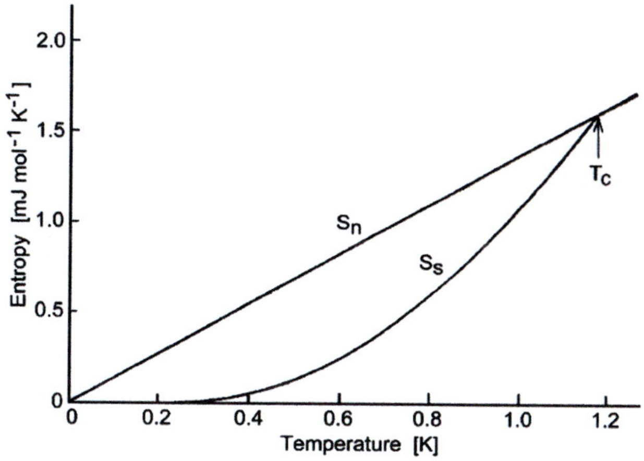

Figure 8.13. Entropy vs temperature plot for Al in superconducting) and normal states ().

Source: C. Buzzea, Creative Commons License.

Figure 8.12b displays the free energy fluctuation for both superconducting and normal states in relation to applied field. The superconducting phase's free energy is less than the normal state's up to the critical field. There is equality of free energies at the critical field. The superconducting phase's free energy increases beyond the critical field. The free energy of a normal metal remains constant regardless of the applied field (Kivelson et al., 2001).

The definition of entropy is . Thus, using (8.6) yields:

$$\left(\frac{\partial G_n}{\partial T}\right)_p - \left(\frac{\partial G_s}{\partial T}\right)_p = -(1/8\pi)2H_c\left(\frac{\partial H_c}{\partial T}\right)_p \tag{8.6}$$

or,

$$S_n - S_s = -\frac{H_c}{4\pi}\frac{\partial H_c}{\partial T}. \tag{8.7}$$

– is positive since p is always a negative quantity. In the presence of a magnetic field, the entropy of the normal state is higher than that of the superconducting state. The findings of the aluminum experiment are shown in Figure 8.13. The reduced entropy of the superconducting state suggests that the electrons are in a highly organized condition (Tian & Chen, 2016).

The latent heat:

$$Q = T(S_n - S_s) = -T\frac{H_c}{4\pi}\frac{\partial H_c}{\partial T}. \tag{8.8}$$

At . Therefore, at the transition temperature, the latent heat is zero.

8.4.1. Specific Heat

The equation for determining the specific heat per unit volume, C, is expressed as . The equation calculates the disparity in particular temperatures between the normal and superconducting states (Zeller & Pohl, 1971). By using equation (8.8), we acquire:

$$C_n - C_s = -\left(\frac{T}{4\pi}\right)\left[H_c\frac{\partial^2 H_c}{\partial T^2} + \left(\frac{\partial H_c}{\partial T}\right)^2\right]. \tag{8.9}$$

At the temperature of transition, T = Tc and Hc = 0 are found. Then (8.9) becomes:

$$C_s - C_n = \Delta C = \left(\frac{T_c}{4\pi}\right)\left(\frac{\partial H_c}{\partial T}\right)^2. \tag{8.10}$$

Equation (8.10) is commonly referred to as Rutger's law and has been experimentally confirmed. By combining equation (8.10) with equation (8.1), we obtain:

$$C_s - C_n = \left(\frac{3}{2\pi}\right)\left(\frac{H_0^2}{T_c}\right)\left[\left(\frac{T}{T_c}\right)^3 - \frac{1}{3}\left(\frac{T}{T_c}\right)\right]. \tag{8.11}$$

Both and are composed of two constituents, the lattice constituent (indexed as l) and the electronic constituent (indexed as e). The expressions for and can be written as and . Assuming that the lattice components remain constant in both the normal state and the superconducting state, equation (8.11) can be expressed in an alternative manner (Butland & Maddison, 1973).

$$C_s - C_n = C_{es} - C_{en} = \left(\frac{3}{2\pi}\right)\left(\frac{H_0^2}{T_c}\right)\left[\left(\frac{T}{T_c}\right)^3 - \frac{1}{3}\left(\frac{T}{T_c}\right)\right]. \tag{8.12}$$

It is known from the free electron theory that:

$$C_{en} = \gamma T. \tag{8.13}$$

By comparing equations (8.12) and (8.13), we obtain:

$$\gamma = \left(\frac{1}{2\pi}\right)\left(\frac{H_0^2}{T_c^2}\right). \tag{8.14}$$

Clearly, represents the term in equation (8.12). Therefore, we have:

$$C_{es} = 3\gamma T_c\left(\frac{T}{T_c}\right)^3. \tag{8.15}$$

The experimental findings are in accordance with equation (8.15), with the exception of extremely low temperatures (Lu et al., 2001).

An exponential drop in the electronic contribution has been found in superconducting phase studies carried out at very low temperatures.

Considering this, (8.15) can be adjusted as:

$$\frac{C_{es}}{\gamma T_c} = a \exp(-bT_c/T) \tag{8.16}$$

where: the values of a and b remain constant. Additionally, equation (8.16) can be expressed as:

$$\frac{C_{es}}{\gamma T_c} = a \exp(-\Delta/k_B T) \tag{8.17}$$

where: Δ is known as the energy gap parameter; this parameter will be covered in a different section.

The absence of latent heat and a quick change in specific heat at the transition point characterize a second order transition. One can observe that the superconducting transition satisfies these features and is categorized as a second order transition by looking at equations (8.8) and (8.10) (Rajan, 1983).

We have explored various theoretical aspects of specific heats in both the superconducting and normal states. Now, let's analyze how these theoretical findings align with the empirical data:

- Figure 8.14 compares the experimental findings on tin from Keesom and Laer's study with ΔC from (8.9). Values from (8.9) are shown by the dashed curve. It is evident that the deal is fair (Campbell et al., 1991);
- Data on the specific heat of gallium in its normal and superconducting phases are displayed in Figure 8.15. A 200 Gauss magnetic field was applied to get data for the normal condition below. At , there is an abrupt decrease in specific heat (Moler et al., 1997).

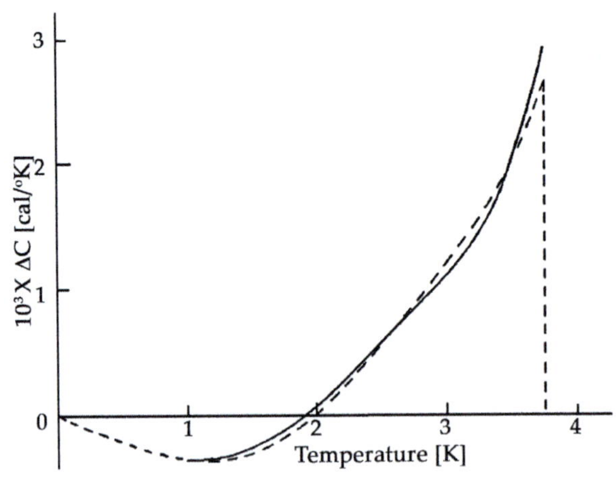

Figure 8.14. Illustration of temperature fluctuation in ΔC for tin; dashed curve from (8.9) and smooth curve representing experimental data.

Source: J.G. Bednorz, Creative Commons License.

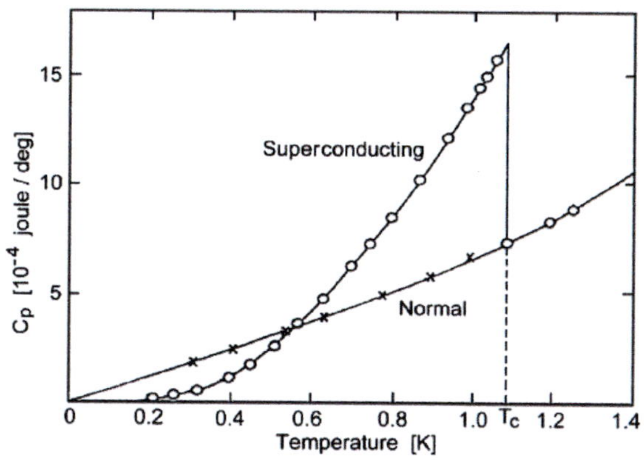

Figure 8.15. Specific heat () of gallium varies with temperature in both its superconducting and normal states; data for the normal state below T_c were acquired by using a magnetic field.

Source: N. Phillips, Creative Commons License.

8.4.2. Energy Gap

A superconductor's electron energy spectrum features a gap. This is the energy superconducting electrons need to become regular electrons. Figure 8.16 displays the location of the energy gap () in relation to the full band and the Fermi energy (). Compared to the energy gap in a semiconductor, the energy gap is far smaller. The electron-lattice interaction is where the energy gap in a semiconductor originates. In relation to the Fermi gas, this contact maintains the electrons' ordered state (Sham & Schlüter, 1983).

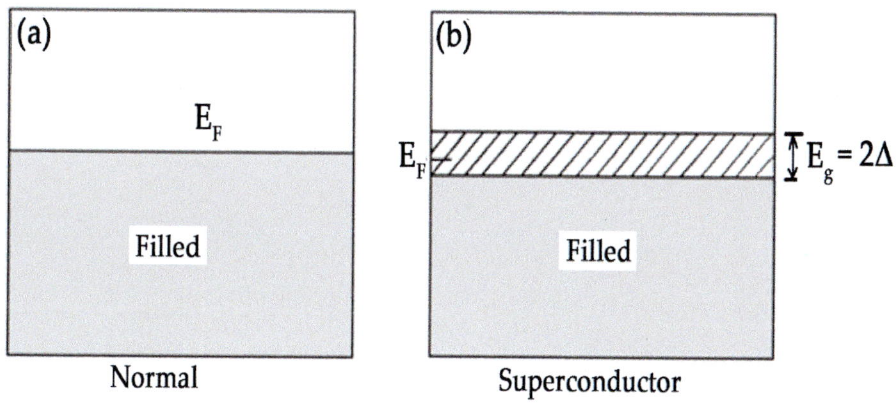

Figure 8.16. Illustration of (a) Conduction band filled in a typical metal; (b) the superconductor energy gap at the Fermi level EF.

Source: C.J. Gorter, Creative Commons License.

Table 8.5. Energy Gap Values some Superconductors by Different Methods

Superconductor		in units of	
	From Infrared Absorption	From Specific Heat	From Electron Tunneling
Sn	3.6	3.6	4
In	4.1	3.9	4
Pb	4.1		4.2
Hg (☒)			4.6
Nb			3.8
AI			3.4
Zn			3.2
V			3.4
Ta			3.6
Cd			3.2

Source: J. Bardeen, Creative Commons License.

A superconductor's energy gap is temperature-dependent. As the temperature gets closer to the critical temperature , it drops to zero. The expression for Ces (8.17) introduces the energy gap parameter Δ, which is double the value of the energy gap (Varshni, 1967).

From the experimental values of , the energy gap may be calculated. Other phenomena that will be covered later, such as quantum tunneling and radiation absorption, can also be used to identify it. Table 8.5 provides the values of E_g for a number of superconducting elements (Panish & Casey, 1969).

8.5. BCS THEORY OF SUPERCONDUCTIVITY

A thorough microscopic theory (henceforth referred to as the BCS theory) was proposed by Bardeen, Cooper, and Schreiffer over fifty years after superconductivity was discovered. The idea uses complex mathematical methods that are beyond of our scope. Thus, a qualitative approach is given.

Separately, Frohlich and Bardeen proposed that electrons in a phonon field might absorb and emit phonons, which could lead to an attractive interaction that produces superconductivity. Frohlich proposed that the electron-phonon interaction contributes to superconductivity while working on research on polaron states in ionic crystals (Wada & Fukuda, 1959)

Furthermore, he proposed that for isotopes of a given element should be proportional to the Debye temperature if electron-phonon interactions are significant. should change as since is known to be proportional to , where M is the isotopic mass. The ratio was in fact verified by later experimental data on the of isotopes of various elements, supporting Frolich's contention that electron-phonon interaction has a crucial part in superconductivity (Komnik & Thorwart, 2016). Since superconductivity, like superfluidity, is a quantum phenomenon, magnetic flux should also be quantized in units, according to the London equation. Further research conducted by Fairbank and Deaver revealed that the flux is quantized, but interestingly, the quantized unit of flux is found to be only half of its original value, which is . This suggests that the charge of the superconducting carriers is –2e, consistent with the idea of Cooper pairs. Cooper suggested that states containing attractively interacting pairs of electrons could exist.

In addition, two electrons in motion inside a phonon field can interact with each other through the exchange of virtual phonons. According to the uncertainty principle, a virtual phonon is one that is absorbed before it has had time to be detected. Cooper pairs are the name for these electrons. It is notable that superconductivity is not explained by a boson condensation process in the BCS theory. Superconducting carriers are made up of electron pairs, while regular carriers are made up of unpaired electrons (Sigrist & Ueda, 1991).

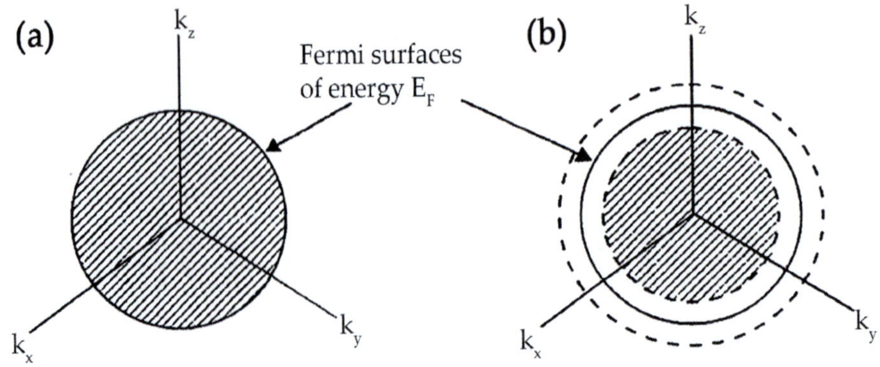

Figure 8.17. Illustration of (a) Fermi level of a typical metal where all states are occupied below EF; (b) the BCS ground state.

Source: B.S. Deaver, Creative Commons License.

Due to attractive contact, the BCS ground state characterizes every electron in the superconductor at zero temperature as opposed to the occupied Fermi level of the normal state (Figure 8.17a, b).

Cooper pairs make up the ground state of the BCS system; if one electron state is unoccupied, the other must be as well. When some electron states just inside the regular Fermi surface get occupied while others just inside the surface remain unoccupied, this deviates from the normal state (see Figure 8.18) (Asomba,1995).

Electron pairs occupy the orbitals in a BCS state. There must also be a filled orbital with a wave vector –k and a spin down ↑ if one with a wave vector k and a spin up ↑ exists. In the event where ↑ is empty, –↓ is also empty. Cooper pairs are zero-spin bosons with several boson-like properties.

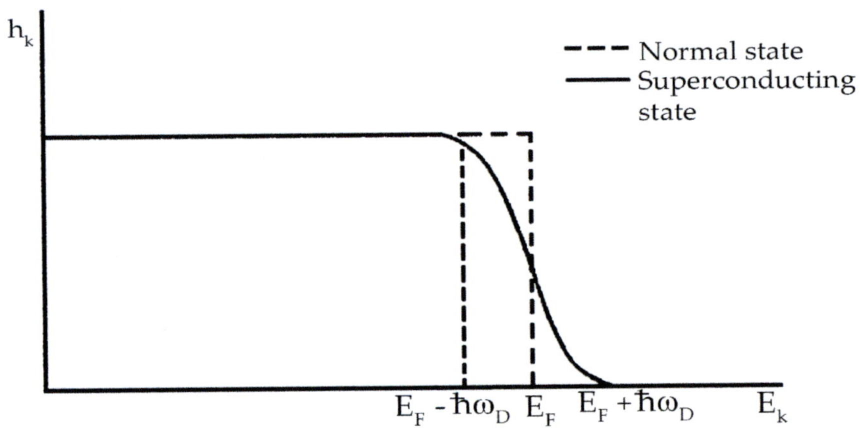

Figure 8.18. Illustration of electrons in a superconducting BCS state have an occupation probability of ; some electrons that were previously in normal states and were in the energy range to are now in the energy states between to .

Source: I. Giaever, Creative Commons License.

Superconductivity is facilitated by the interaction between electrons and phonons, which helps reduce the repulsion between electrons caused by Coulombic forces. A Cooper pair, consisting of an electron with spin up and spin down (, can transition to any vacant position of another pair, consisting of an electron with spin up (↑) and an electron with spin down (↓) through the exchange of a virtual phonon with momentum q. A Cooper pair is a composite particle formed by the binding of two electrons. The energy gap represents the difference in energy between unpaired and paired state. It is equal to twice the energy gap parameter Δ, expressed as = 2Δ. To generate two quasi-particle excitations, a Cooper pair must be disrupted, which necessitates an energy (Figure 8.19).

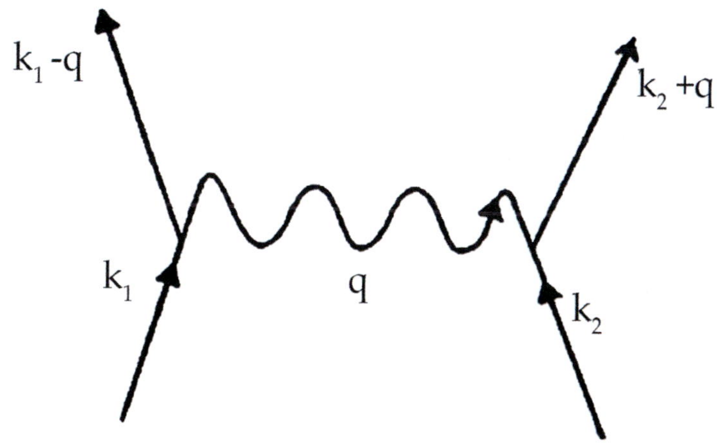

Figure 8.19. Illustration of virtual phonons exchange between electrons.

Source: K.A. Muller, Creative Commons License.

For 2Δ at 0 K, the BCS theory provides the following connection.

$$\frac{2\Delta(0)}{k_B T_c} = 3.52. \quad (8.18)$$

This relationship is universally applicable as it does not involve any phenomenological variables. The majority of them agree with (8.18), however there is a notable divergence for Pb and Hg. An energy gap based on temperature is predicted by the BCS theory. The temperature variation around is provided by:

$$\frac{\Delta(T)}{\Delta_0} = 1.76\left(1 - \frac{T}{T_c}\right)^{1/2}. \quad (8.19)$$

Equation (8.19) implies that as , . An alternative reading of (8.19) would be that as temperature rises, the proportion of superconducting electrons diminishes until all electrons are normal electrons at T = Tc. Figure 8.20 displays experimental results on the temperature fluctuation of Δ (T) for a few superconductors. The BCS theory (8.19) is in line with these findings:

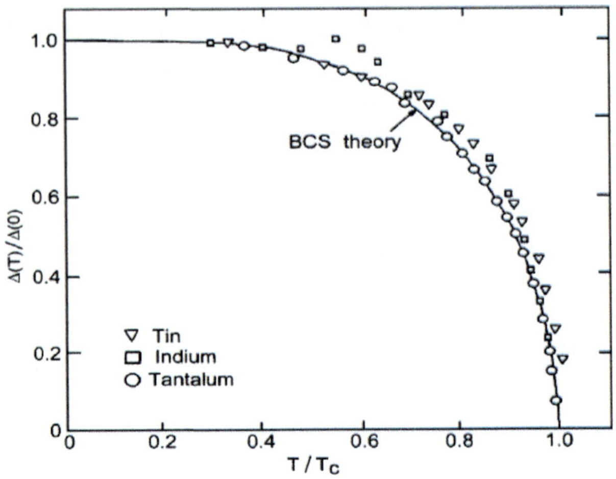

Figure 8.20. Temperature variation of the superconducting energy gap; a comparison between BCS theory and experimental data.

Source: R.H. Willes, Creative Commons License.

The BCS theory's most significant concept is a bound state of two electrons, called Cooper pairs. Many characteristics of superconductors, such as the spectroscopic absorption, specific heat, Meissner effect, critical field, and even the London equation, can be observed as a result of this theory (Gor'kov, 1960).

8.6. APPLICATIONS OF SUPERCONDUCTIVITY

Superconductivity has a wide range of uses. Only a select few are mentioned in the following sections:

8.6.1. Power Transmission

Superconductors are more advantageous than regular metals for large scale applications due to their low resistance, resulting in less power loss. However, the expenses associated with producing low temperatures and managing superconducting materials provide conventional materials a greater benefit. However, the utilization of HTS materials reduces the need for iron cores in generators, motors, and transformers, leading to a decrease in size, weight, and losses associated with iron cores. Additionally, HTS materials belong to the category of type II superconductors that exhibit exceptional tolerance to high magnetic fields. These characteristics render them exceptionally advantageous (Sachtjen et al., 2000).

8.6.2. Levitation

Superconductors exhibit the phenomenon of levitation. As a result, high-speed trains that use magnetic levitation now use liquid helium. The MAG-LEV trains have been tested in Japan, Germany, and the USA (Andrade et al., 2018).

8.6.3. Squids

Superconducting materials were first used in electronics as instruments, and they still are, primarily because of their exceptional sensitivity in magnetic fields. Numerous small-scale applications with low current requirements have resulted from this. These are SQUID magnetometers, infrared detectors, and Josephson junction devices (Welty & Martinis, 1991).

8.6.4. Bio Magnetism

Bio magnetism is a significant sector in which superconductor devices are applied. HTS materials are used to create Josephson junctions, which are used in medical diagnostic procedures including evaluating heart and brain function. Magnetic resonance imaging is the term for this method (MRI). Strong magnetic fields generated from superconductors are applied to the human body, forcing the hydrogen atoms found in water and fat molecules to take magnetic field energy (Williamson & Kaufman, 1981).

Then, at a frequency that a computer can detect and graphically show, they release this energy. Although MRI technology was developed in the middle of the 1940s, the first human body MRI was not carried out until 1977. Almost five hours were needed to create a single photograph. These days, the process only takes a few seconds (Bahadur & Giri, 2003).

8.6.5. Satellite Communication

Superconductors are not only incredibly sensitive magnetic field detectors, but they are also excellent at detecting very weak electromagnetic signals, such as those coming from space. High sensitivity SQUID detectors are found at many radio astronomy observatories throughout the world. HTS thin films have been produced into filters and resonators, among other passive microwave devices, for use in space technology (Rahmat & Densmore, 2014).

The latest breakthrough in satellite communication is the development of a "all-digital receiver" on a compact superconducting niobium chip, measuring just one centimeter in size. Around 11,000 Josephson junctions are arranged on the device to create superconducting Rapid Signal Flux Quantum (RSFQ) circuits (Sadek & Aissa, 2012).

ACTIVITY 8.1

Objective: Demonstrate magnetic levitation using a superconductor and observe the Meissner effect.

Materials Needed:

- Superconducting material (e.g., YBCO disk);
- Small neodymium magnet;
- Liquid nitrogen (or another cryogenic cooling agent);
- Safety gloves and goggles (for handling liquid nitrogen).

Instructions:

a. Prepare the Superconductor:

- Place the superconducting material (YBCO disk) on a stable surface.

b. Cool the Superconductor:

- Pour liquid nitrogen over the superconductor to cool it down below its critical temperature. Ensure gloves and goggles are worn for safety.

c. Place the Magnet:

- Position the small neodymium magnet above the cooled superconductor, ensuring it is not in direct contact.

d. Observe Magnetic Levitation:

- Witness the magnet hovering above the superconductor due to the expulsion of magnetic flux (Meissner effect). This demonstrates the zero friction and perfect diamagnetism properties of superconductors.

SUMMARY

- The chapter on superconductivity begins with a historical overview of its discovery, tracing key milestones in the understanding of this phenomenon. It explores the conditions under which superconductivity occurs, detailing the types of materials and temperatures involved.
- Superconductors exhibit distinctive properties such as zero electrical resistance and perfect diamagnetism, which are essential to their practical applications. The chapter covers thermal properties, including how superconductors behave at different temperatures, and magnetic properties, such as their response to magnetic fields and the Meissner effect.
- Thermodynamic aspects of the transition into and out of the superconducting state are discussed, alongside specific heat changes and the concept of an energy gap in electronic states at low temperatures.
- The BCS theory is presented as the foundational explanation for superconductivity in conventional materials. Practical applications of superconductivity are explored, including its role in power transmission efficiency, levitation technologies, SQUIDs for precise magnetic field measurements, biomagnetism research, and advancements in satellite communication systems.

REVIEW QUESTIONS

1. What are the key properties that distinguish superconductors from regular conductors?
2. Describe the conditions under which superconductivity occurs in materials.
3. Explain the Meissner effect and its significance in superconductors.
4. How does the BCS theory explain the phenomenon of superconductivity?
5. What are the thermal properties of superconductors and how do they change with temperature?
6. Discuss the magnetic properties of superconductors and their applications.
7. What is the energy gap in superconductors, and how does it relate to their behavior at low temperatures?
8. Name and describe three practical applications of superconductivity.
9. What are SQUIDs and how are they used in scientific research?

MULTIPLE CHOICE QUESTIONS

1. What is a defining property of superconductors?

a. High electrical resistance

b. Low thermal conductivity

c. Zero electrical resistance

d. Strong magnetic susceptibility

2. **Which theory explains conventional superconductivity?**
    a. General Relativity
    b. Quantum Mechanics
    c. BCS Theory
    d. Special Relativity
3. **The Meissner effect in superconductors refers to:**
    a. Zero electrical resistance
    b. Perfect diamagnetism
    c. High thermal conductivity
    d. Low specific heat
4. **Superconductors typically lose their properties above:**
    a. 0 K
    b. 77 K
    c. 273 K
    d. 1000 K
5. **What is the most common type of material exhibiting superconductivity at higher temperatures?**
    a. Aluminum
    b. Lead
    c. Copper
    d. Yttrium Barium Copper Oxide (YBCO)
6. **SQUIDs are used primarily for:**
    a. Levitation
    b. Measuring magnetic fields
    c. Power transmission
    d. Studying thermal properties

Answers to Multiple Questions

1. (c); 2. (c); 3. (b); 4. (b); 5. (d); 6. (b)

REFERENCES

1. Abrikosov, A. A. (1957). The magnetic properties of superconducting alloys. Journal of Physics and Chemistry of Solids, 2(3), 199–208.
2. Andrade, M. A., Pérez, N., & Adamowski, J. C. (2018). Review of progress in acoustic levitation. Brazilian Journal of Physics, 48, 190–213.

3. Araujo-Moreira, F. M., Lisboa-Filho, P. N., Lanfredi, A. J., Ortiz, W. A., Zanetti, S. M., Leite, E. R., ... & Venkatesan, V. (2000). The search for superconductivity in PrBa2Cu3O7– δ. Physica C: Superconductivity, 341, 413–416.
4. Asomba, G. C. (1995). On BCS theory of superconductivity in systems with overlapping bands I. Simple overlap. Physica C: Superconductivity, 244(3–4), 271–281.
5. Bahadur, D., & Giri, J. (2003). Biomaterials and magnetism. Sadhana, 28, 639–656.
6. Baral, P. C., & Rout, G. C. (2012). A modified BCS theory of heavy fermion superconductivity. Indian Journal of Physics, 86, 431–437.
7. Bednorz, J. G., Müller, K. A., & Takashige, M. (1987). Superconductivity in alkaline earth-substituted La2CuO4-y. Science, 236(4797), 73–75.
8. Bergfield, J. P., Solis, M. A., & Stafford, C. A. (2010). Giant thermoelectric effect from transmission supernodes. ACS nano, 4(9), 5314–5320.
9. Bouchard, L. S., Acosta, V. M., Bauch, E., & Budker, D. (2011). Detection of the Meissner effect with a diamond magnetometer. New Journal of Physics, 13(2), 025017.
10. Bristow, K. L. (2002). 5.3 Thermal Conductivity. Methods of Soil Analysis: Part 4 Physical Methods, 5, 1209–1226.
11. Butland, A. T. D., & Maddison, R. J. (1973). The specific heat of graphite: an evaluation of measurements. Journal of Nuclear Materials, 49(1), 45–56.
12. Campbell, G. S., Calissendorff, C., & Williams, J. H. (1991). Probe for measuring soil specific heat using a heat-pulse method. Soil Science Society of America Journal, 55(1), 291–293.
13. Carrington, A., Giannetta, R. W., Kim, J. T., & Giapintzakis, J. (1999). Absence of nonlinear Meissner effect in YBa 2 Cu 3 O 6.95. Physical Review B, 59(22), R14173.
14. Carruthers, P. (1961). Theory of thermal conductivity of solids at low temperatures. Reviews of Modern Physics, 33(1), 92.
15. Cava, R. J., Batlogg, B., Espinosa, G. P., Ramirez, A. P., Krajewski, J. J., Peck Jr, W. F., ... & Cooper, A. S. (1989). Superconductivity at 3.5 K in BaPb0. 75Sb0. 25O3: why is T c so low?. Nature, 339(6222), 291–293.
16. Dahl, P. F. (1984). Kamerlingh onnes and the discovery of superconductivity: The leyden years, 1911–1914. Historical Studies in the Physical Sciences, 15(1), 1–37.
17. Emery, N., Hérold, C., d'Astuto, M., Garcia, V., Bellin, C., Marêché, J. F., ... & Loupias, G. (2005). Superconductivity of bulk CaC 6. Physical review letters, 95(8), 087003.
18. Eremets, M. I., & Drozdov, A. P. (2016). High-temperature conventional superconductivity. Physics-Uspekhi, 59(11), 1154.
19. Essén, H., & Fiolhais, M. C. (2012). Meissner effect, diamagnetism, and classical physics—a review. American Journal of Physics, 80(2), 164–169.
20. Finnemore, D. K., Ostenson, J. E., Bud'ko, S. L., Lapertot, G., & Canfield, P. C. (2001). Thermodynamic and transport properties of superconducting Mg 10 B 2. Physical Review Letters, 86(11), 2420.

21. Gabovich, A. M., & Kuznetsov, V. I. (2013). What do we mean when using the acronym 'BCS'? The Bardeen–Cooper–Schrieffer theory of superconductivity. European Journal of Physics, 34(2), 371.
22. Galambos, T. H., Ronetti, F., Hetényi, B., Loss, D., & Klinovaja, J. (2022). Crossed Andreev reflection in spin-polarized chiral edge states due to the Meissner effect. Physical Review B, 106(7), 075410.
23. Geim, A. K., Dubonos, S. V., Lok, J. G. S., Henini, M., & Maan, J. C. (1998). Paramagnetic Meissner effect in small superconductors. Nature, 396(6707), 144–146.
24. Gor'kov, L. P. (1960). The critical supercooling field in superconductivity theory. Soviet Physics JETP, 10, 593–599.
25. Gorsky, A., Shifman, M., & Yung, A. (2005). Non-Abelian Meissner effect in Yang-Mills theories at weak coupling. Physical Review D—Particles, Fields, Gravitation, and Cosmology, 71(4), 045010.
26. Hirsch, J. E. (2009). BCS theory of superconductivity: it is time to question its validity. Physica Scripta, 80(3), 035702.
27. Hirsch, J. E., & Marsiglio, F. (2021). Meissner effect in nonstandard superconductors. Physica C: Superconductivity and its applications, 587, 1353896.
28. Kamiya, K., Takeuchi, T., Kabeya, N., Wada, N., Ishimasa, T., Ochiai, A., ... & Sato, N. K. (2018). Discovery of superconductivity in quasicrystal. Nature communications, 9(1), 154.
29. Kivelson, S. A., Aeppli, G., & Emery, V. J. (2001). Thermodynamics of the interplay between magnetism and high-temperature superconductivity. Proceedings of the National Academy of Sciences, 98(21), 11903–11907.
30. Komissarov, S. S., & McKinney, J. C. (2007). The 'meissner effect'and the blandford—znajek mechanism in conductive black hole magnetospheres. Monthly Notices of the Royal Astronomical Society: Letters, 377(1), L49-L53.
31. Komnik, A., & Thorwart, M. (2016). BCS theory of driven superconductivity. The European Physical Journal B, 89, 1–5.
32. Konczykowski, M., Burlachkov, L. I., Yeshurun, Y., & Holtzberg, F. (1991). Evidence for surface barriers and their effect on irreversibility and lower-critical-field measurements in Y-Ba-Cu-O crystals. Physical Review B, 43(16), 13707.
33. Kozhevnikov, V. (2021). Meissner effect: History of development and novel aspects. Journal of Superconductivity and Novel Magnetism, 34(8), 1979–2009.
34. Kumar, P., Hall, D., & Goodrich, R. G. (1999). Thermodynamics of the superconducting phase transition in Ba 0.6 K 0.4 BiO 3. Physical review letters, 82(22), 4532.
35. Larbalestier, D., & Canfield, P. C. (2011). Superconductivity at 100—Where we've been and where we're going. Mrs Bulletin, 36(8), 590–593.
36. Liu, R. H., Luo, X. G., Zhang, M., Wang, A. F., Ying, J. J., Wang, X. F., ... & Chen, X. H. (2011). Coexistence of superconductivity and antiferromagnetism in single crystals A0. 8Fe2− ySe2 (A= K, Rb, Cs, Tl/K and Tl/Rb): evidence from magnetization and resistivity. Europhysics Letters, 94(2), 27008.

37. Lu, L., Yi, W., & Zhang, D. L. (2001). 3ω method for specific heat and thermal conductivity measurements. Review of scientific instruments, 72(7), 2996–3003.

38. Magnetic Materials, 22(2), 129–201.

39. Makinson, R. E. B. (1938, July). The thermal conductivity of metals. In Mathematical Proceedings of the Cambridge Philosophical Society (Vol. 34, No. 3, pp. 474–497). Cambridge University Press.

40. Matthias, B. T. (1957). Chapter v superconductivity in the periodic system. In Progress in low temperature physics (Vol. 2, pp. 138–150). Elsevier.

41. Moler, K. A., Sisson, D. L., Urbach, J. S., Beasley, M. R., Kapitulnik, A., Baar, D. J., ... & Hardy, W. N. (1997). Specific heat of YBa 2 Cu 3 O 7– δ. Physical Review B, 55(6), 3954.

42. Panish, M. B., & Casey, H. C. (1969). Temperature dependence of the energy gap in GaAs and GaP. Journal of Applied Physics, 40(1), 163–167.

43. Putilin, S. N., Antipov, E. V., Chmaissem, O., & Marezio, M. (1993). Superconductivity at 94 K in HgBa2Cu04+ δ. Nature, 362(6417), 226–228.

44. Rahmat-Samii, Y., & Densmore, A. C. (2014). Technology trends and challenges of antennas for satellite communication systems. IEEE Transactions on Antennas and Propagation, 63(4), 1191–1204.

45. Rajan, V. T. (1983). Magnetic susceptibility and specific heat of the Coqblin-Schrieffer model. Physical review letters, 51(4), 308.

46. Rao, N. A. H. K., Garland, J. C., & Tanner, D. B. (1984). Thermodynamic transition of small superconducting particles. Physical Review B, 29(3), 1214.

47. Sachtjen, M. L., Carreras, B. A., & Lynch, V. E. (2000). Disturbances in a power transmission system. Physical Review E, 61(5), 4877.

48. Sadek, M., & Aissa, S. (2012). Personal satellite communication: Technologies and challenges. IEEE wireless communications, 19(6), 28–35.

49. Samokhin, K. V. (2004). Magnetic properties of superconductors with strong spin-orbit coupling. Physical Review B—Condensed Matter and Materials Physics, 70(10), 104521.

50. Schmidt, H., Weber, M., & Braun, H. F. (1995). Occurrence and composition dependence of superconductivity and magnetism in HoNi2B2C. Physica C: Superconductivity, 246(1–2), 177–185.

51. Sham, L. J., & Schlüter, M. (1983). Density-functional theory of the energy gap. Physical review letters, 51(20), 1888.

52. Sigrist, M., & Ueda, K. (1991). Phenomenological theory of unconventional superconductivity. Reviews of Modern physics, 63(2), 239.

53. Simon, A. (1997). Superconductivity and chemistry. Angewandte Chemie International Edition in English, 36(17), 1788–1806.

54. Slocombe, D. R., Kuznetsov, V. L., Grochala, W., Williams, R. J., & Edwards, P. P. (2015). Superconductivity in transition metals. Philosophical Transactions of the Royal Society A: Mathematical, Physical and Engineering Sciences, 373(2037), 20140476.

55. Suzuki, T., Ishiguro, K., Mori, Y., & Sekido, T. (2005). Dual Meissner effect and magnetic displacement currents. Physical review letters, 94(13), 132001.
56. Tanigaki, K., Ebbesen, T. W., Saito, S., Mizuki, J., Tsai, J. S., Kubo, Y., & Kuroshima, S. (1991). Superconductivity at 33 K in CsxRbyC60. Nature, 352(6332), 222–223.
57. Tian, W., & Chen, H. (2016). Insight into the mechanical, thermodynamics and superconductor properties of NbRuB via first-principles calculation. Scientific reports, 6(1), 19055.
58. Toberer, E. S., Baranowski, L. L., & Dames, C. (2012). Advances in thermal conductivity. Annual Review of Materials Research, 42(1), 179–209.
59. Van Delft, D. (2012). History and significance of the discovery of superconductivity by Kamerlingh Onnes in 1911. Physica C: Superconductivity, 479, 30–35.
60. Varshni, Y. P. (1967). Temperature dependence of the energy gap in semiconductors. physica, 34(1), 149–154.
61. Wada, Y., & Fukuda, N. (1959). On the Theory of Superconductivity. Progress of Theoretical Physics, 22(6), 775–806.
62. Weihrich, R., Matar, S. F., Anusca, I., Pielnhofer, F., Peter, P., Bachhuber, F., & Eyert, V. (2011). Palladium site ordering and the occurrence of superconductivity in Bi2Pd3Se2− xSx. Journal of Solid State Chemistry, 184(4), 797–804.
63. Welty, R. P., & Martinis, J. M. (1991). A series array of DC SQUIDs. IEEE Transactions on Magnetics, 27(2), 2924–2926.
64. Werthamer, N. R., Helfand, E. F., & Hohenberg, P. C. (1966). Temperature and purity dependence of the superconducting critical field, H c 2. III. Electron spin and spin-orbit effects. Physical Review, 147(1), 295.
65. Zeller, R. C., & Pohl, R. O. (1971). Thermal conductivity and specific heat of noncrystalline solids. Physical Review B, 4(6), 2029.

Index

Printed and bound by CPI Group (UK) Ltd, Croydon, CR0 4YY
18/09/2026
14965064-0001